Understanding
Laser Accidents

Understanding Laser Accidents

Edited by
Ken Barat

CRC Press
Taylor & Francis Group
Boca Raton London New York

CRC Press is an imprint of the
Taylor & Francis Group, an **informa** business

CRC Press
Taylor & Francis Group
6000 Broken Sound Parkway NW, Suite 300
Boca Raton, FL 33487-2742

First issued in paperback 2023

© 2019 by Taylor & Francis Group, LLC
CRC Press is an imprint of Taylor & Francis Group, an Informa business

No claim to original U.S. Government works

ISBN-13: 978-1-138-04845-4 (hbk)
ISBN-13: 978-1-03-265253-5 (pbk)
ISBN-13: 978-1-315-11438-5 (ebk)

DOI: 10.1201/b22255

Publisher's Note
The publisher has gone to great lengths to ensure the quality of this reprint but points out that some imperfections in the original copies may be apparent.

Visit the Taylor & Francis Web site at
http://www.taylorandfrancis.com

and the CRC Press Web site at
http://www.crcpress.com

Book dedications are an opportunity to say thank you. I owe a great deal of my success and joy of science to the many laser users I have met over the years. The author Issacs Asimov did a lot to instill my wonder in science. The Bay Area Laser Safety Officers Group, which had a 13-year run, was also a major influence. The tolerance of my wife (Pat) and children (Emily and Leah) freed me to reach my goals. Now I have a granddaughter to make proud (Margo). So, thank you all from deep within my heart.

Contents

Editor

Ken Barat is the former Laser Safety Officer (LSO) for Lawrence Berkeley National Lab and the National Ignition Facility Directorate. Presently, he is providing laser safety consulting under the name of Laser Safety Solutions. He is the author of several texts on laser safety as well as numerous articles and presentations worldwide.

He is an OSA distinguished speaker, Fellow of the Laser Institute of America (LIA), and a senior member of IEEE and SPIE. He was the chair and organizer of the first seven LSO workshops and an ANSI committee member and chair. For laser questions, he is the "Ask the Expert" for the Health Physics Society. He is a former LIA board member and laser safety instructor for several laser professional societies and institutions, a Rockwell award winner, and among the first class of Certified Laser Safety Officers.

Contributors

Larry Johnson started in fiber optics in 1977. He has been an active part of this dynamic industry. He is director and founder of The Light Brigade, Inc. His specialities include: fiber-optic design, planning, installation, maintenance, and operations.

Karen Kelley, CIH, CLSO, is the Director of Research Safety at Cornell University. She has a BS in Mathematics, an MS in Environmental Health, and over 20 years of experience in research safety, industrial hygiene, and laser safety in both industry and academia. Karen previously served as the Laser Safety Officer at the University of Pennsylvania and the University of Maryland. A focus of her work over the past few years has been on safety culture in academic research.

Roberta McHatton specializes in laser entertainment and display applications. As an active member of the leading laser standard organizations ANSI and SAE, Roberta provides 15 plus years of experience to high-profile high-risk laser applications by helping her clients successfully implement superior laser safety programs worldwide. A vast knowledge of regulatory requirements and Roberta's network of contacts makes her work valuable to this specialized industry.

Patrick Murphy holds a BA degree in Laser Art and Technology and an MBA degree. He began working on laser airspace issues in the mid-1990s with the SAE G10-T Laser Safety Hazards Committee. In this capacity, he helped write the regulations and forms used by the U.S. Federal Aviation Administration to evaluate laser usage in airspace. He received an Award of Recognition from SAE G10-T for this work. He has presented papers at the International Laser Safety Conference in 1997, 2009, 2011, 2015, and 2017 on the topics of laser/aircraft safety and audience-scanned laser shows. He is currently co-chair of the SAE G10-OL Operational Laser Committee, helping to draft a document on pilot education and protective eyewear, including running tests on pilots with lasers and bright lights in cockpits. Since 2006, he has been executive director of the International Laser Display Association.

Randy Paura, P. Eng, CLSO, President of Dynamic Laser Solutions Inc, University of Waterloo, Ontario, Canada, Professional Engineer of Ontario (since 1991), is an industrial laser materials processing consultant that includes laser system solutions (automation and process). He provides industrial laser safety review training and audits and conducts laser process audits and optimization.

Randy Paura is a certified laser safety officer, Vice-Chair ANSI Z136.9 "Safe Use of Lasers in Manufacturing Environments." He is a member of AWS C7C American Welding Society, Subcommittee on Laser Welding. Randy is experienced in all aspects of the design and build process of special purpose machinery. His specialities include: innovative laser materials processing solutions, and process/system reorganization for optimized/minimal floor space requirements.

Tekla A. Staley, CIH, CSP, CLSO, is a senior health and safety engineer at Idaho National Laboratory (INL). She has a BS in chemistry with a minor in business management from the College of Idaho and over 30 years of experience in the field of industrial hygiene and safety. Tekla has been employed at INL since 1990 with assignments at the Reactor Technologies Complex, Test Area North, Idaho Nuclear Technologies and Engineering Center, and Specific Manufacturing Capabilities, and is experienced in semiconductor manufacturing safety, asbestos building inspections and abatement, microscopic analysis of asbestos, and lead abatement. She is responsible for sitewide asbestos and laser safety programs; and she was assigned to the Three Mile Island Fuel Drying/Storage and the Cask Dismantlement projects. Tekla has consulted with the U.S. Army on proposed methods for destruction of VX nerve agent and assists DOE-ID in providing independent industrial hygiene assistance at U.S. Army contract facilities in the United States. She contracts with the Laser Institute of America to co-instruct Laser Safety Officer classes. Tekla is a member of ANSI Z136 committees for Safe Use of Lasers in Research, Development, or Testing and Safe Use of Lasers in Manufacturing Environments and the Energy Facility Contractors Group Laser Safety Task Group. She has presented information on laser incident investigations and hazardous energy control methods at the International Laser Safety Conference and Department of Energy Laser Safety Officer Workshops. Additionally, she uses her health and safety experience in the agricultural industry while owning/operating a 420-acre beef cattle and grass hay production ranch with her husband in Butte County, Idaho.

Introduction: Short and to the Point

The best learning tool is to see what one's peers have done. In this case, it would be ways of not making the same mistakes.

The first laser was invented in 1960, and the first documented laser accident in 1964. The goal of this book is not to run through an endless list of gory accidents, but rather to address the topic of laser accidents in a holistic means. The book covers the breath of laser accidents, from being prepared, to culture, where to find accident information and so much more.

I have collected chapters from what I consider the best people in their respective fields. I hope the reader will take the information to heart and take steps to make sure their stories are not included in volume two of such a text.

I greatly appreciate all who gave up their time to contribute to this text. I hope you, as the reader, will feel the same way.

Ken Barat

1 Why Accidents Happen

Ken Barat

CONTENTS

Why do laser accidents happen? Everyone has an opinion. The answer changes with the laser setting, and with the most popular safety philosophy of that year. The two answers that I think will endure the passage of time are: no negative consequences to bad behavior and poor to no mentoring. These two items are the ones I would like to explore in this chapter.

BAD BEHAVIOR AND NO NEGATIVE CONSEQUENCES

The above title seems to contradict itself. But once one thinks about it, the words take on a clear meaning and highlight a major contributor to not only laser accidents, but all accidents, excluding equipment failure. So, let me clearly state that this chapter deals with human actions. Other laser accident causes will be discussed later.

Statistics show us that most laser accidents happen to experienced laser users. The question of why seems to be a natural one. An article I recently read seems to offer one answer.*

It started with the question: Why do people, including myself, drive over the posted speed limit? Well, the answer is that the behavior has occurred hundreds to thousands of times without negative consequences. If one got a ticket each time they exceeded the speed limit, that behavior would change and change rapidly.

The same applies to experimental laser users—not checking for stray reflections, peaking under/over eyewear, not wearing eyewear—each time we violate good laser safety practice and get away with it, that action reenforces the poor practice. I am not advocating for injuries, just stating a fact. If one was hit in the eye each time they looked under their eyewear, no doubt, the practice would stop.

Looking at the list below, which highlights actions that are usually credited with the cause of most laser accidents, how many of these get a contributing factor from

---

* David Mallard, "The Human Factor in Safety & Operation," *EHS today*, June 2016, www.EHStoday.com.

the fact that users have done them repeatedly, before something happened to them, causing them to say "Wow, that was a close call," or "Stupid, do not do that again."

1. **Unanticipated eye exposure during alignment**
   It is a universal law that beam manipulation can produce unexpected reflections. It is this activity that separates R&D laser work from all other laser use settings, where beam manipulation is an exception or is the only service activity not an almost daily activity. Good alignment practices are, therefore, critical to preventing laser accidents.

2. **Misaligned optics and upwardly directed beams**
   Unless one is working with vertical breadboards, lasers are thought of in the horizontal plane. Hence, any time beams are directed upward to change beam height, a risk is present. So, one needs to be aware of this condition and take steps to protect themselves and others.

3. **Available laser protective eyewear not used**
   Yes, this happens all the time. It all traces back to a list of reasons and excuses for the user not to use their eyewear: the eyewear is a problem, poor visibility, excessive weight, poor fit, and so on. The other reason is just being lazy/over confident or the user is so familiar with the activity that they don't feel that the eyewear is needed.

4. **Equipment malfunction**
   Many times, this happens due to a lack of a preventive maintenance program or consideration of such. Environmental conditions can also add to this, such as dirt, blocked ventilation holes, poor handling, or storage.

5. **Improper methods of handling high voltage**
   This can be as simple as the lack of grounding or improper grounding, frayed wires … the list on electrical safety goes on.

6. **Intentional exposure of unprotected personnel**
   Yes, there have been a few such cases. So, follow the golden rule: treat others as you would have them treat you.

7. **Operators unfamiliar with laser equipment**
   This goes back to improper on-the-job training/mentoring. Sometimes, it is the individual wishing to appear more experienced or knowledgeable to others they are working with. In some cases, it is having worked with similar equipment before and therefore believing they can figure it out or it is all the same. This is very common in industrial settings, counting on "skill of the craft."

8. **Lack of protection for non-beam hazards**
   While we focus on laser hazards, they are very commonly not the greatest risk to the laser user. As I have said before, laser eyewear does very little if one is electrocuted. Same goes for chemical burns or suffocation from gases.

9. **Improper restoration of equipment following service**
   Anytime that service is performed on equipment, there should be a verification that interlock by passes have been removed and safety systems are back functioning as expected.

10. **Laser protective eyewear worn not appropriate for laser in use**
    This is most common in laser use areas that have multiple pairs of eyewear for different laser use. Therefore, awareness of what the activity is and checking one's eyewear is critical.
11. **Unanticipated eye/skin exposure during laser usage**
    This is, most commonly, putting one's hands in the beam path, an exposure from a stray reflection, or a beam escaping a broken fiber.
12. **Inhalation of LGAC and/or viewing laser-generated plasmas**
    This comes from not using a smoke evacuator or a poorly functioning one. The worst case would be not using any means to capture the plume or being unaware of its risk to one's health.
13. **Fires resulting from the ignition of materials**
    This comes from not being aware of the irradiance of the beam and the combustion potential of materials being used. A simple example is when I have seen people use Post-It Notes as beam blocks for stray reflections.
14. **Eye or skin injury of photochemical origin**
    This is from unblocked UV scatter, which is not felt while one is being overexposed, as well as the effect showing up or being felt even 10–12 hours later.
15. **Failure to follow Standard Operating Procedures (SOPs)**
    If the SOP is instructional, it should be followed, but many times, users do not follow an SOP as they feel there's no value other than to make the Safety Department feel good or to satisfy safety rules.
16. **Introduction of foreign materials (pages of loose paper, paper clips, falling items, or objects)**
    We never want unintended items to fall into the beam path, this goes back to housekeeping or set-up procedures.
17. **Modification of the beam path**
    There is nothing wrong with modifying beam paths as needed, but one must realize any such action presents the opportunity for stray reflections, which can cause injury if not blocked or mitigated.
18. **Poor communications**
    Communication is the key to safety; each person working on the lab must be aware or keep aware of the actions of others. Especially if one person has removed a beam block, moved an optic, and so on. This extends to entering information into research logs so people can read what changes may exist before they start their work.
19. **Turning to look at the source of bright light detected by peripheral vision**
    Unexpended bright light or flashes are never a good sign and how one decides to look or react to them can lead to more trouble.

POSSIBLE FUTURE: VIRTUAL REALITY

While not commonly used, there are a few facilities and software that allow a user to wear virtual reality (VR) and explore a laser room, including the feel of performing alignment. It is clear as VR becomes more accessible that this approach could be a common training aid for laser users and laser technicians. Keep in mind that the flight simulator had a significant and documented impact in reducing aircraft crashes, because the flight crew could experience and practice responses to unanticipated challenges, such as a bird sucked into an engine or single engine failure. The VR approach is an excellent bridge to the next section in this chapter.

ON THE JOB TRAINING/MENTORING

Despite efforts to improve pilot performance, crashes due to pilot error remained at 65% for more than 50 years. That changed in 1990, when flight simulators for pilot training became standard, which provided a tool designed to provide experimental learning in a safe and controlled setting. Since then, crashes due to pilot error have declined by more than 54%, now fewer than 3.4 defects per 1 million opportunities.

For us in the laser community, the days of flight simulators are not that far off. I have already encountered virtual reality units that practice laser alignment.

But for most laser users, it is mentoring (on-the-job training), being taught good practices, and seeing them practiced that makes the difference, so they become the expected mode of operation, not the exception. The majority of incidents occur in activities that are perceived safe, which is when one is most likely to cheat on the way it should be done. Perceived safety comes from violating good practice with no consequences. We cannot accept risk-taking during laser work if not for our own protection but for those around us. Laser eye injuries are like a ball in a roulette wheel—one does not know where it will land or how severe the injury might be. Get smart, be safe, I am now stepping down from the soapbox.

The greatest risk of on-the-job training (OJT) or mentoring is that of mentoring bad habits. Sometimes, this is presented to new users as tricks of the trade. Unfortunately, they do not have the experience to understand the basics or risks of these so-called tricks/shortcuts present.

Bad habits are not only deliberately learned activities, but more commonly observed bad habits. When the senior person does not wear laser protective eyewear during alignment or beam manipulation, the novice sees this and gets the message that this is the way things are done.

If unexplained, they may miss that a shutter has been put in place, so the risk has been removed or other mitigation is in place, if that is the case.

Laser work in a research lab has elements of art built into it. How the artist handles the brush will yield different effects on the canvas. How one learns to handle optics and mounts will affect the quality of work and time it takes to set-up work. This includes items like fingerprints on optics, setting the initial screw settings on optical mounts, or labeling mounts... the list goes on.

RECOMMENDATION

OJT needs to be documented. No one is given an expensive laboratory system and told to "GO for it." There is always some orientation period. That period should involve instruction, observations, a question period, and finally approval. The trainer and trainee both need to sign a form that they are satisfied the trainee can work on their own. This can be done in a step-by-step process or an all-in-one approach. By documenting this agreement, all parties go on record that a certain level of competence has been reached.

TRAINER, WHAT ARE YOUR RESPONSIBILITIES?

One of the key tasks the OJT instructor must perform when meeting with a new trainee is to review the entire goals of the OJT with the trainee. The OJT instructor should emphasize the expectations of the training, including the skills the trainee will need to acquire to be fully qualified. The instructor will need to explain how the training will be conducted, how the trainee will be able to successfully achieve each task, and limits on the trainee and how they will change over the OJT process.

The trainer needs to explain what they are doing, why they are doing the tasks, and the safety precautions that must be considered to do the job safely. A common trainer problem is to overwhelm the trainee with details on the first day.

TRAINEE, WHAT ARE YOUR RESPONSIBILITIES?

ASK Questions, don't be afraid to look uninformed, so again ASK questions.

A trainee should ask questions for clarity when in doubt and keep notes and look for additional supporting self-study materials.

The steps of the process need to include the following items:

1. *Demonstrate how to perform a task.* The OJT instructor should select a simple task and demonstrate how to perform the task correctly, explaining why the task is to be performed, its importance, and the impact to the operation of the part or system if not done correctly.
2. *Allow the trainee to perform part of the task.* In this phase of the instruction, the trainer and trainee are interacting together, with the trainer coaching as necessary.
3. *Allow the trainee to perform the entire task* with coaching as necessary from the OJT trainer. Depending on the task complexity, this may occur several times before the task is mastered.
4. *Evaluate the trainee's performance of a task.* Observe the worker performing the entire task without supervision. When the worker can perform the task without supervision, they are considered trained. Trainee OJT record can be constructed so that each series of tasks can be signed off as an accomplishment.
5. *Sign off the training package*. To confirm their competency, the worker should be allowed to perform the complete task without active supervision. At this point, the trainee's training for that particular task will have been completed, and the instructor can sign off.
6. *Keep records.* The final signature page should be filed in a secure place. This may be in the form of physical storage or a digital storage system. The training record should be maintained for as long as the organization deems necessary. For some organizations, the minimum time for maintaining training records is 5 years.

What Makes Good Coaching?

Do you praise, curse, or just roll one's eyes? One of the most difficult jobs the OJT instructor faces is providing trainees with feedback about their progress. Giving the impression that the trainee is progressing well when they need more coaching is counterproductive and dishonest to the trainee. On the other hand, being too direct or frank about a trainee's progress can easily be mistaken for criticism, especially if there are no established criteria used to evaluate the trainee's performance. Feedback provides trainees with an idea of how well they are performing. Therefore, one should:

- Provide an immediate and complete answer to the task item after the trainee's completion; all parts of the answer or answers should be provided. If alternative methods for completing the tasks are acceptable, each should be included.
- Promptly give the trainee practical means to better understand a task in which they underperformed. For example, go over the task and the procedures to which the task relates and then have the trainee practice under your supervision.
- Provide guidance for remediation. The purpose of feedback is to help trainees learn the material. Therefore, OJT should be designed so that the trainee is led to restudy the information they failed to recall, recognize, or perform.

2 Laser Safety Terms

Ken Barat

CONTENTS

As in all disciplines, there is a list of essential term; in other words, a language. The same is true for laser safety and laser technology. From a laser safety perspective, there are about 10 critical terms and many more, depending on the depth one dwells into laser standards and laser use.

TEN DEFINITIONS YOU MUST KNOW

1. **Maximum Permissible Exposure (MPE)**
   The level of laser radiation to which an unprotected person may be exposed without adverse biological changes in the eye or skin. One way to think of MPE is as a posted speed limit. Driving up to that limit, one should not receive a speeding limit. The faster one drives over the limit, the more likely an accident or ticket will happen.
2. **Nominal Hazard Zone (NHZ)**
   The space or zone within which the level of the direct, reflected, or scattered radiation may exceed the applicable MPE. Exposure levels outside of the boundary of the NHZ will not cause injury. Think of it as the area within a circle or oval.
3. **Nominal Ocular Hazard Zone (NOHZ)**
   The distance along the axis of the unobstructed beam from a laser, fiber end, or connector to the human eye. Think of it as the diameter of the NHZ. Beyond that diameter or axis, the exposure is below the MPE for eyes or skin.
4. **Optical Density (OD)**
   The logarithm ratio of the incident beam to the transmitted beam of a wavelength through a filter material. Think of OD as SPF for sunblock. The higher the OD, the less photons penetrate the filter. OD is usually found on laser eyewear lens or protective windows.

$$OD = \log_{10}(1/\tau\lambda)$$

   where $\tau\lambda$ is the transmittance at the wavelength of interest. *Symbol:* OD, $D(\lambda)$, $D\lambda$.

5. **Irradiance**

   Output of the beam divided by the area of the beam diameter. Another way to express it is the radiant power incident per unit area upon a surface, expressed in watts-per-centimeter-squared ($W \cdot cm^{-2}$). *Symbol: E.*

6. **Access Exposure Limit (AEL)**

   This term is used for laser product classification. For each laser hazard classification, there is an associated output measurement. Per the CDRH, AEL means: the magnitude of accessible laser or collateral radiation of a specific wavelength and emission duration at a particular point as measured by CDRH requirements, according to paragraph (e) of US 49 CFR part 1040. Accessible laser or collateral radiation is radiation to which human access is possible,

7. **Continuous Wave (CW)**

   Within the laser standards, CW is a laser operating with a continuous output for a period ≥ 0.25 s. In many ways, one can think of it as a flashlight with continuous output.

8. **Pulsed Beam**

   A laser which delivers its energy in the form of a single pulse or a train of pulses. In the laser standard, the duration of a pulse is less than 0.25 s. Once a pulse laser's reputation rate is over 25 thousand pulses per second, it can be considered to act like a CW laser.

9. **Diffuse Reflection**

   A beam bouncing off a surface where the surface irregularities are greater than the size of the wavelength. This is the safest type of reflection to be exposed to (Figure 2.1).

10. **Specular Reflection**

    A beam bouncing off a surface where the surface irregularities are smaller than the wavelength, therefore the reflected energy can be close to the primary beam (Figure 2.2).

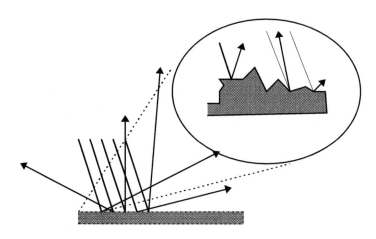

FIGURE 2.1   Diffuse reflection surface features.

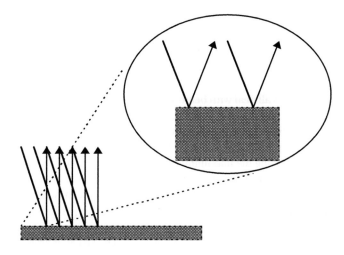

FIGURE 2.2   Specular reflection surface features.

THE MOST MISUNDERSTOOD TERMS AND DEFINITIONS

1. **Protective Housing vs. Enclosure**

   The official ANSI Z136 definition of protective housing is: An enclosure that surrounds the laser or laser system and prevents access to laser radiation above the applicable MPE. The aperture through which the useful beam is emitted is not part of the protective housing. The protective housing limits access to other associated radiant energy emissions and to electrical hazards associated with components and terminals, and may enclose associated optics and a workstation.

   For most of us, it means the cover that covers the internal optics of a commercial laser. These PHs are interlocked so the laser shuts down if removed.

   There is no official Z136.1 definition of an enclosure; there is one for an enclosed laser. So the general consensus is an enclosure would be a box covering a laser system. Therefore, a protective housing would have engineering controls per the classification of the laser, while many times, an enclosure is added by the user.

2. **Class 1 Product and Embedded Laser System**

   A Class 1 laser product is a classification where during normal operation there is no hazard from the laser source. Think of a laser printer. It is a CDRH or IEC term. While an embedded laser is an ANSI term for a higher-class laser inside a piece of equipment where, once again, there is no exposure potential, but may not have all the engineering controls as would be required of a CDRH Certified product.

3. **Aversion Response vs. Blink Reflex**

   For years, blink reflex and aversion response were considered comparable if not interchangeable. Research studies have demonstrated that there is a difference. An aversion response is the closure of the eyelid, eye movement, pupillary constriction, or movement of the head to avoid an exposure to a bright light stimulant. For a laser user, the aversion response to an exposure from a bright, visible, laser source is assumed to limit the exposure of a specific retinal area to 0.25 s or less. This seems to be universal, while blink response is an involuntary closure of the eyelids after exposure to a bright light which can be slower than the 0.25 seconds of the aversion response.

4. **$1/e^2$: These values all deal with beam width or sometimes termed beam waist**
   $1/e^2$ is the beam diameter as the distance between diametrically opposed points in that cross section of a beam where the power per unit area is $1/e$ (0.368) times that of the peak power per unit area. This is the beam diameter definition that is used for computing the maximum permissible exposure to a laser beam.

   $1/e$ is a mathematical way to express beam diameter. The distance between diametrically opposed points in that cross section of a beam where the power per unit area is $1/e$ (0.368) times that of the peak power per unit area.

5. **Laser Control Area**
   For IEC and ANSI compliance it is defined as "The space or area where the occupancy and activity of those within is subject to control and supervision for the purpose of protection from laser radiation hazards." While the ANSI laser research standard Z136.8 breaks indoor laser control areas into laser use locations and divides into the following:

   a. **Unrestricted location:** Access is not limited. By default, no laser radiation hazards exist (Class 1 or embedded laser system), and these locations can be occupied by the general public, visitors, and spectators without implementing administrative or engineering control measures, and personal protective equipment (PPE).

      Example: A public area with embedded laser systems, or displays, such as a museum exhibit. A biotechnology lab using Class 1 laser products, that is, gene sequencer cell sorters.

   b. **Restricted location:** Access is granted for authorized people and limited for the general public through administrative and engineering control measures. Laser radiation hazards at Class 3B levels or greater may be present and control measures are required. Administrative controls include posted warning signs, attending training, and following established SOPs for laser system(s). Engineering controls include access control measures such as lockable doors, barriers, undefeatable interlocks, and curtains to prevent laser radiation from leaving the restricted location.

      Example: A research laboratory or fabrication area containing Class 3B and/or Class 4 lasers.

   c. **Controlled location:** Access, occupancy, and activities of people within are subject to strict control and supervision. By inference, controlled locations are restricted locations with laser radiation hazards at Class 4 with additional control measures specified by the laser operator, the LSO, and the employer management.

      Example: An R&D area with positive access control and video surveillance, a clean room.

   d. **Exclusion location:** Occupancy by people is possible, but is denied by the LSO during the operation of the laser system.

Example: A free electron laser target/experimental room or beam path, or an area with an open kilowatt beam path.

e. **Inaccessible location:** Occupancy is not possible due to its dimensions.
   Example: An enclosed beam path on an optical table or open path of a laser scanning confocal microscope.

6. **Average Power vs. Peak Power**
   Average power is a term used with pulsed lasers. It is the energy per pulse times the number of pulses in a second. The results are in Watts. Its value will always be lower than the peak power result.

   Peak Power is also a term used with pulse lasers. It is the energy per pulse (just like average power) but divided by the pulse duration. Its value is in Joules/cm^2.

7. **Near Field vs. Far Field**
   The terms' meanings are very much like the names imply. Far field is the region/area, far from the aperture of the laser. The near field *of* a laser beam is understood to be the region around the beam waist, the focus. The far field concerns the beam profile far from the beam waist.

dB AND WATTS

Decibel is the common unit to express power gain or loss. The term is commonly used in telecommunications (Table 2.1).

TABLE 2.1
dB conversion to Watts

Power (dBm)	Power (mW)
−40 dBm	0.0001 mW
−30 dBm	0.001 mW
−20 dBm	0.01 mW
−10 dBm	0.1 mW
0 dBm	1 mW
1 dBm	1.2589 mW
2 dBm	1.5849 mW
3 dBm	1.9953 mW
4 dBm	2.5119 mW
5 dBm	3.1628 mW
6 dBm	3.9811 mW
7 dBm	5.0119 mW
8 dBm	6.3096 mW
9 dBm	7.9433 mW
10 dBm	10 mW
20 dBm	100 mW
30 dBm	1000 mW
40 dBm	10000 mW
50 dBm	100000 mW

AN ASSORTMENT OF USEFUL LASER SAFETY
AND RELATED DEFINITIONS

The terms and definitions listed below relate to how one would use them when considering or applying laser safety applications or questions. The symbols used are as they would be found in various laser standards.

absorption: The loss of light as it passes through a material.

adsorption: Transformation of radiant energy to a different form of energy by interaction with matter.

active medium: A medium in which stimulated emission will take place at a given wavelength.

acuity: A measure of the eye's ability to resolve detail.

administrative control measure: Control measures incorporating administrative means [e.g., training, safety approvals, LSO designation, and Standard Operating Procedures (SOP)] to mitigate the potential hazards associated with laser use.

aperture: An opening or window through which radiation passes.

attenuation: The decrease in the radiant flux as it passes through an absorbing or scattering medium.

authorized personnel: Individuals approved by management to install, operate, or service laser equipment. Needs to be an individual that has completed the requirements to work with the most common Class 3B and/or Class 4 lasers.

beam expander: An optical device that increases beam diameter while decreasing beam divergence (spread). In its simplest form it consists of two lenses, the first to diverge the beam and the second to re-collimate it. Also called an upcollimator.

beam splitter: An optical device using controlled reflection to produce two beams from a single incident beam.

Brewster windows: The transmissive end (or both ends) of the laser tube, made of transparent optical material and set at Brewster's angle in gas lasers to achieve zero reflective loss for one axis of plane polarized light. They are nonstandard on industrial lasers, but some polarizing element must be used if a polarized output is desired.

carcinogen: An agent potentially capable of causing cancer.

coagulation: The process of congealing by an increase in viscosity characterized by a condensation of material from a liquid to a gelatinous or solid state.

coherent: A light beam is said to be coherent when the electric vector at any point in it is related to that at any other point by a definite, continuous function.

collateral radiation: Any electromagnetic radiation, except laser radiation, emitted by a laser or laser system which is physically necessary for its operation.

collecting optics: Lenses or optical instruments having magnification and thereby producing an increase in energy or power density. Such devices may include telescopes, binoculars, microscopes, or loupes.

collimated beam: Effectively, a "parallel" beam of light with very low divergence or convergence.

divergence (φ): For the purposes of this standard, divergence is taken as the plane angle projection of the cone that includes 1–1/e (i.e., 63.2%) of the total radiant energy or power. The value of the divergence is expressed in radians or milliradians.

duty cycle: Ratio of total "on" duration to total exposure duration for a repetitively pulsed laser.

energy: The capacity for doing work. Energy content is commonly used to characterize the output from pulsed lasers, and is generally expressed in joules (J).

erythema: is the medical term for redness of the skin due to congestion of the capillaries.

extended source: A source of optical radiation with an angular subtense at the cornea larger than α_{min}.

eye-safe laser: A Class 1 laser product. Because of the frequent misuse of the term "eye-safe wavelength" to mean "retina-safe," (e.g., at 1.5–1.6 µm) and "eye-safe laser" to refer to lasers emitting outside the retinal-hazard region in this spectral region, the term "eye-safe" can be a misnomer. Hence, the use of "eye-safe laser" is discouraged.

fail-safe interlock: An interlock where the failure of a single mechanical or electrical component of the interlock will cause the system to go into, or remain in, a safe mode.

femtoseconds: 10^{-15} seconds. 1 fs = 0.000,000,000,000,001 secs.

fiber optics: A system of flexible quartz or glass fibers that use total internal reflection (TIR) to pass light through thousands of glancing (total internal) reflections.

flashlamp: A tube typically filled with krypton or xenon. Produces a high intensity white light in short duration pulses.

fluorescence: The emission of light of a particular wavelength resulting from absorption of energy typically from light of shorter wavelengths.

focal length: The distance, measured in centimeters, from the secondary nodal point of a lens to the secondary focal point. For a thin lens imaging a distant source, the focal length is the distance between the lens and the focal point.

focal point: The point toward which radiation converges or from which radiation diverges or appears to diverge.

gain: Another term for amplification.

gas laser: A type of laser in which the laser action takes place in a gas medium.

goggles: Laser protective eyewear intended to fit the face surrounding the eyes in order to shield the eyes from certain hazards. Goggles are often designed to be worn over prescription eyeglasses to seal against the face to provide peripheral protection for the eyes from laser radiation.

half-power point: The value on either the leading or trailing edge of a laser pulse at which the power is one-half of its maximum value.

hertz (Hz): The unit which expresses the frequency of a periodic oscillation in cycles per second.

image: The optical reproduction of an object, produced by a lens or mirror. A typical positive lens converges rays to form a "real" image which can be photographed. A negative lens spreads rays to form a "virtual" image which can't be projected.

incident light: A ray of light that falls on the surface of a lens or any other object. The "angle of incidence" is the angle made by the ray with a perpendicular (normal) to the surface.

infrared: The region of the electromagnetic spectrum between the long-wavelength extreme of the visible spectrum (about 0.7 μm) and the shortest microwaves (about 1 mm).

infrared radiation: Electromagnetic radiation with wavelengths which lie within the range 0.7 μm to 1 mm.

infrared radiation (IR): is electromagnetic radiation with wavelengths that lie within the range 0.7–1 mm.

installation: Placement and connection of laser equipment at the appropriate site to enable intended operation.

intrabeam viewing: The viewing condition whereby the eye is exposed to all or part of a laser beam.

ionizing radiation: Electromagnetic radiation having a sufficiently large photon energy to directly ionize atomic or molecular systems with a single quantum event.

iris: The circular pigmented membrane which lies behind the cornea of the human eye. The iris is perforated by the pupil.

irradiance: Radiant power incident per unit area upon a surface, expressed in watts-per-square-centimeter ($W \cdot cm^{-2}$). Synonym: *power density.*

joule (J): A unit of energy. 1 joule = 1 watt · second. (W · s).

KTP (Potassium Titanyl Phosphate): A crystal used to change the wavelength of an Nd:YAG laser from 1060 nm (infrared) to 532 nm (green).

Lambertian surface: An ideal diffuse surface whose emitted or reflected radiance is independent of the viewing angle.

laser: A device that produces radiant energy predominantly by stimulated emission. Laser radiation may be highly coherent temporally, or spatially, or both. An acronym for **L**ight **A**mplification by **S**timulated **E**mission of **R**adiation.

laser barrier: A device used to block or attenuate incident direct or diffuse laser radiation. Laser barriers are frequently used during times of service to the laser system when it is desirable to establish a boundary for a temporary (or permanent) laser-controlled area.

laser diode: A laser employing a forward-biased semiconductor junction as the active medium. Synonyms: *injection laser; semiconductor laser.*

laser pointer: A Class 2 or Class 3a laser product that is usually handheld, which emits a low-divergence visible beam of less than 5 milliwatts and is intended for designating specific objects or images during discussions, lectures, or presentations as well as for the aiming of firearms or other visual targeting practice.

laser safety officer(LSO): One who has authority to monitor and enforce the control of laser hazards and effect the knowledgeable evaluation and control of laser hazards.

laser system: An assembly of electrical, mechanical, and optical components which includes a laser.

lesion: An abnormal change in the structure of an organ or part due to injury or disease.

macula: The small uniquely pigmented specialized area of the retina of the eye, which, in normal individuals, is predominantly employed for acute central vision (i.e., area of best visual acuity).

magnified viewing: Viewing a small object through an optic that increases the apparent object size. This type of optical system can make a diverging laser beam more hazardous (e.g., using a magnifying optic to view an optical fiber with a laser beam emitted).

maintenance: Performance of those adjustments or procedures (specified in user information provided by the manufacturer with the laser or laser system), which are to be performed by the user to ensure the intended performance of the product.

manometer (nm): A unit of length in the International System of Units (SI) equal to one billionth of a meter. Abbreviated nm—a measure of length. One nm equals 10^{-9} meter, and is the usual measure of light wavelengths. Visible light ranges from about 400 nm in the purple to about 760 nm in the deep red.

meter: A unit of length in the international system of units; currently defined as the length of a path traversed in vacuum by light during a period of 1/299792458 seconds. Typically, the meter is subdivided into the following units:

$$\text{centimeter (cm)} = 10^{-2}\ m$$
$$\text{millimeter (mm)} = 10^{-3}\ m$$
$$\text{micrometer (}\mu m\text{)} = 10^{-6}\ m$$
$$\text{nanometer (nm)} = 10^{-9}\ m$$

mode: A term used to describe how the power of a laser beam is geometrically distributed across the cross section of the beam. Also used to describe the operating mode of a laser such as continuous or pulsed.

mode locked: A method of producing laser pulses in which short pulses (approximately 10^{-12} second) are produced and emitted in bursts or a continuous train.

modulation: The ability to superimpose an external signal on the output beam of the laser as a control.

monochromatic light: Theoretically, light consisting of just one wavelength. No light is absolutely single frequency since it will have some bandwidth. Lasers provide the narrowest of bandwidths that can be achieved.

multimode: Laser emission at several closely spaced frequencies.

nanosecond: One billionth (10^{-9}) of a second. Longer than a picosecond or femtosecond, but shorter than a microsecond. Associated with Q-switched lasers.

Nd:glass laser: A solid-state laser of neodymium: glass offering high power in short pulses. An Nd doped glass rod used as a laser medium to produce 1064 nm light.

Nd:YAG laser Neodymium:Yttrium Aluminum Garnet: A synthetic crystal used as a laser medium.

Neodymium (Nd): The rare earth element that is the active element in Nd:YAG lasers and Nd:glass lasers.

non-beam hazard: A class of hazards that result from factors other than direct human exposure to a laser beam.

operation: The performance of the laser or laser system over the full range of its intended functions (normal operation).

optical cavity: (Resonator). Space between the laser mirrors where lasing action occurs.

optical fiber: A filament of quartz or other optical material capable of transmitting light along its length by multiple internal reflection and emitting it at the end.

optical pumping: The excitation of the lasing medium by the application of light rather than electrical discharge.

optical radiation: Ultraviolet, visible, and infrared radiation (0.35–1.4 nm) that falls in the region of transmittance of the human eye.

optically aided viewing: Viewing with a telescopic or magnifying optic. Under certain circumstances, viewing with an optical aide can increase the hazard from a laser beam. (See telescopic viewing or magnified viewing.)

optically pumped lasers: A type of laser that derives energy from another light source such as a xenon or krypton flashlamp or other laser source.

output coupler: Partially reflective mirror in laser cavity which allows emission of laser light.

output power: The energy per second measured in watts emitted from the laser in the form of coherent light.

photochemical effect: An effect (e.g., biological effect) produced by a chemical action brought about by the absorption of photons by molecules that directly alter the molecule. For example, one photon of sufficient energy can alter a single molecule. Such effects are generally important in the shorter-visible and ultraviolet regions of the optical spectrum. The threshold radiant exposure is constant over a wide range of exposure durations ("the Bunsen-Roscoe Law").

photosensitizers: Substances which increase the sensitivity of a material to irradiation by electromagnetic energy.

plasma radiation: Black-body radiation generated by luminescence of matter in a laser-generated plume.

protective eyewear (laser): Equipment such as spectacles, face shields, goggles, eye shields, spectacles, and visors intended to provide eye protection from laser beam hazards.

pulse duration: The duration of a laser pulse, usually measured as the time interval between the half-power points on the leading and trailing edges of the pulse.

pulse-repetition frequency (PRF): The number of pulses occurring per second, expressed in hertz.

pulsed laser: A laser which delivers its energy in the form of a single pulse or a train of pulses. In this standard, the duration of a pulse <0.25 s.

pump: To excite the lasing medium. (See optical pumping or pumping.)

pumped medium: Energized laser medium.

pumping: Addition of energy (thermal, electrical, or optical) into the atomic population of the laser medium, necessary to produce a state of population inversion.

pupil: The variable aperture in the iris through which light travels to the interior of the eye.

Q-switch: A device for producing very short ($\sim$10–250 ns), intense laser pulses by enhancing the storage and dumping of electronic energy in and out of the lasing medium, respectively.

radian (rad): A unit of angular measure equal to the angle subtended at the center of a circle by an arc whose length is equal to the radius of the circle. 1 radian $\sim$57.3 degrees; 2π radians = 360 degrees.

radiance: Radiant flux or power output per unit solid angle per unit area expressed in watts-per-centimeter squared-per-steradian ($W \cdot cm^{-2} \cdot sr^{-1}$).

radiant energy: Energy emitted, transferred, or received in the form of radiation. Unit: joules (J).

radiant exposure: Surface density of the radiant energy received, expressed in units of joules-per-centimeter squared ($J \cdot cm^{-2}$).

radiant power: Power emitted, transferred, or received in the form of radiation, expressed in watts (W). Synonym: *radiant flux.*

refraction: The bending of a beam of light in transmission through an interface between two dissimilar media or in a medium whose refractive index is a continuous function of position (graded index medium).

refractive index (of a medium): Denoted by n, the ratio of the velocity of light in vacuum to the phase velocity in the medium. Synonym: *index of refraction.*

repetitive pulse laser: A laser with multiple pulses of radiant energy occurring in a sequence.

resonator: The mirrors (or reflectors) making up the laser cavity, including the laser rod or tube. The mirrors reflect light back and forth to build up amplification.

retina: The sensory membrane that receives the incident image formed by the cornea and lens of the human eye. The retina lines the inside of the eye.

retinal hazard region: Optical radiation with wavelengths between 0.4 and 1.4 μm, where the principal hazard is usually to the retina.

safety latch: A mechanical device designed to slow direct entry to a controlled area.

secured enclosure: An enclosure to which casual access is impeded by an appropriate means, e.g., a door secured by a magnetically or electrically operated lock or latch, or by fasteners that need a tool to remove.

service: The performance of those procedures or adjustments described in the manufacturer's service instructions which may affect any aspect of the performance of the laser or laser system.

shall: The word "shall" is to be understood as mandatory.

should: The word "should" is to be understood as advisory.

small source: In this document, a source with an angular subtense at the cornea equal to or less than alpha-min (α_{min}), that is, $\leq$ than 1.5 mrad. This includes all sources formerly referred to as "point sources" and meeting small-source viewing (formerly called point source or intrabeam viewing) conditions.

Small-source viewing: The viewing condition whereby the angular subtense of the source, α_{min}, is equal to or less than the limiting angular subtense, a_{min}.

solid angle: The three-dimensional angular spread at the vertex of a cone measured by the area intercepted by the cone on a unit sphere whose center is the vertex of the cone. Solid angle is expressed in steradians (sr).

standard operating procedure (SOP): Formal written description of the safety and administrative procedures to be followed in performing a specific task.

steradian (sr): The unit of measure for a solid angle. There are 4π steradians about any point in space.

telescopic viewing: Viewing an object from a long distance to increase its visual size. These systems generally collect light through a large aperture, magnifying hazards from large-beam, collimated lasers.

TEM: Abbreviation for Transverse Electromagnetic Modes. Used to designate the cross-sectional shape of the beam. The radial distribution of intensity across a beam as it exits the optical cavity.

TEM00: The lowest order mode possible with a bell-shaped (Gaussian) distribution of light across the laser beam.

thermal effect: An effect brought about by the temperature elevation of a substance (e.g., biological tissue). Photocoagulation of proteins resulting in a thermal burn is an example. The threshold radiant exposure is dependent upon the duration of exposure and heat transfer from the heated area.

threshold limit (TL): In this standard, the term is applied to laser protective eyewear filters, protective windows, and barriers. The TL is an expression of the "resistance factor" for beam penetration of a laser protective device. This is generally related by the Threshold Limit (TL) of the protective device (expressed in $W \cdot cm^{-2}$ or $J \cdot cm^{-2}$). It is the maximum average irradiance (or radiant exposure) at a given beam diameter for which a laser protective device (e.g., filter, window, barrier, and so on) provides adequate beam resistance. Thus, laser exposures delivered on the protective device at or below the TL will limit beam penetration to levels at or below the applicable MPE.

transmission: Passage of radiation through a medium.

transmittance: The ratio of transmitted power to incident power.

tunable dye laser: A laser whose active medium is a liquid dye, pumped by another laser or flashlamps, to produce various colors of light. The color of light may be tuned by adjusting optical tuning elements and/or changing the dye used.

tunable laser: A laser system that can be "tuned" to emit laser light over a continuous range of wavelengths or frequencies.

ultraviolet radiation: Electromagnetic radiation with wavelengths shorter than those of visible radiation; for the purpose of this standard, 0.18 to 0.4 μm.

uncontrolled area: An area where the occupancy and activity of those within is not subject to control and supervision for the purpose of protection from radiation hazards.

viewing window: Visually transparent parts of enclosures that contain laser processes. It may be possible to observe the laser processes through the viewing windows.

visible radiation (light): In this standard, the term is used to describe electromagnetic radiation which can be detected by the human eye. This term is commonly used to describe wavelengths which lie in the range 0.4 to 0.7 μm.

watt (W): The unit of power or radiant flux. 1 watt = 1 joule-per-second.

watt/cm²: A unit of irradiance used in measuring the amount of power per area of absorbing surface, or per area of CW laser beam.

wave: A sinusoidal undulation or vibration; a form of movement by which all radiant electromagnetic energy travels.

wavelength: The length of the light wave, usually measured from crest to crest, which determines its color. Common units of measurement are the micrometer (micron), the nanometer, and (earlier) the Angstrom unit.

window: A piece of glass (or other material) with plane parallel sides which admits light into or through an optical system and excludes dirt and moisture.

work practices: Procedures used to accomplish a task.

YAG (yttrium aluminum garnet): A widely used solid-state crystal that is composed of yttrium and aluminum oxides, which is doped with a small amount of the rare-earth neodymium.

3 Safety Culture and Incident Investigation

Karen Kelley

CONTENTS

The safety culture of an organization plays an important role in incident reporting and incident investigation. While having a strong culture of safety does not preclude an organization from having incidents, its emphasis on a proactive approach to identifying and mitigating safety concerns can reduce the risk of a serious incident. If an incident does occur, organizations with strong safety cultures focus on the opportunity to learn and improve rather than to blame. Conversely, a poor safety culture can be identified as a cause of an incident. The concept of safety culture as a causal factor of an incident is attributed to the International Atomic Energy Agency (IAEA), the organization tasked with investigating the explosion at Chernobyl. In their report, the IAEA identified "poor safety culture" as a contributing cause of the accident.

DEFINING SAFETY CULTURE

There is not a single definition of safety culture, but one commonly referenced definition comes from the U.S. Nuclear Regulatory Commission (NRC). The NRC defines safety culture in their Safety Culture Policy Statement as "the core values and behaviors resulting from a collective commitment by leaders and individuals to emphasize safety over competing goals to ensure protection of people and the environment." The NRC also identifies nine traits of a positive safety culture:

1. **Leadership Safety Values and Actions:** Leaders demonstrate a commitment to safety in their decisions and behaviors.
2. **Problem Identification and Resolution:** Issues potentially impacting safety are promptly identified, fully evaluated, and promptly addressed and corrected commensurate with their significance.

3. **Personal Accountability:** All individuals take personal responsibility for safety.
4. **Work Processes:** The process of planning and controlling work activities is implemented so that safety is maintained.
5. **Continuous Learning:** Opportunities to learn about ways to ensure safety are sought out and implemented.
6. **Environment for Raising Concerns:** A safety conscious work environment is maintained where personnel feel free to raise safety concerns without fear of retaliation, intimidation, harassment, or discrimination.
7. **Effective Safety Communication:** Communications maintain a focus on safety.
8. **Respectful Work Environment:** Trust and respect permeate the organization.
9. **Questioning Attitude:** Individuals avoid complacency and continuously challenge existing conditions and activities in order to identify discrepancies that might result in error or inappropriate action.

FOCUS ON SAFETY CULTURE

Safety culture has long been a focus in industries where the consequences of failure are severe, such as the chemical industry, the nuclear industry, aviation, and health care. More recently, there has been a national focus on safety culture in academic research following several tragic laboratory accidents, including the death of a researcher from a chemical fire in a laboratory at UCLA in 2009, and the serious injury of a graduate student from an explosion in a laboratory at Texas Tech University in 2010. While these incidents were not related to laser use, they exposed systemic gaps in safety culture within academic research communities.

The explosion at Texas Tech University prompted an investigation by the U.S. Chemical Safety Board (CSB), which was the first time the CSB investigated an incident at a university. The CSB investigation found "systemic deficiencies within Texas Tech that contributed to the incident," including the absence of a system to document, track, and communicate lessons learned associated with previous incidents."

These accidents also led to the release of three reports on academic safety culture: *Creating Safety Cultures in Academic Institutions: A Report of the Safety Culture Task Force of the ACS Committee on Chemical Safety*, American Chemical Society (2012); *Safe Science: Promoting a Culture of Safety in Academic Chemical Research*, National Academies of Science (2014); and *A Guide to Implementing a Safety Culture in Our Universities*, Association of Public and Land Grant Universities (2016).

While these reports focus on safety culture in academic research, the findings and recommendations can be broadly applied, and all highlight the importance of near miss and incident reporting, incident investigation and root cause analysis, and learning from incidents. The ACS report recommends that all laboratory incidents be investigated to identify direct, indirect, and root causes. The report states, "Sharing lessons learned through incidents is important in building a strong safety culture." The National Academies of Science and APLU reports recommend that organizations "incorporate non-punitive incident and near-miss reporting as part of their safety cultures," and call on institutions to implement a process to report incidents and near misses and communicate the lessons learned to the campus community.

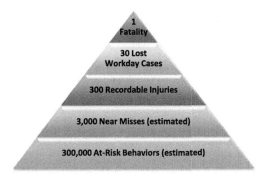

FIGURE 3.1   ConocoPhillips Marine Safety Pyramid, 2003.

THE IMPORTANCE OF NEAR MISS REPORTING

Fortunately, incidents involving serious injuries are not common. For each serious incident, there are many more unsafe acts and conditions, near misses, and minor incidents, and these can often precede a more serious incident. In 2003, ConocoPhillips Marine conducted a study that demonstrated a significant difference between near misses and serious injuries, which is seen in the safety pyramid (Figure 3.1).

Organizations with a strong safety culture support an ongoing process of continuous improvement, which involves identifying unsafe acts and conditions and implementing change proactively to prevent incidents from occurring. As part of this, these organizations often have a formal system for reporting and tracking near misses.

A near miss is an event that had the potential to cause injury or damage if the circumstances were slightly different. An example of a near miss would be a stray beam generated during an alignment procedure entering another area of the lab where personnel were working.

Unfortunately, near misses are often underreported. One barrier to near miss reporting is fear of embarrassment or punishment. According to the National Safety Council, an organization can encourage near miss reporting by establishing a reporting culture that is non-punitive, and anonymous, if desired. An organization can use near miss reporting as a leading indicator, and report back to the employees the positive changes that have been implemented as a result. Another barrier to near miss reporting is a lack of understanding of what a near miss is and how to identify that one has occurred. Many organizations train employees on how to report incidents, and this training can be expanded to include how to identify and report near misses.

INCIDENT CAUSATION

In order to identify and correct the causes of an incident, it is important to understand why incidents happen. Modern incident causation theories, such as system theory or multiple-cause theory, recognize that incidents are not the result of a single malfunction or action, but rather the result of multiple system or organizational failures. James

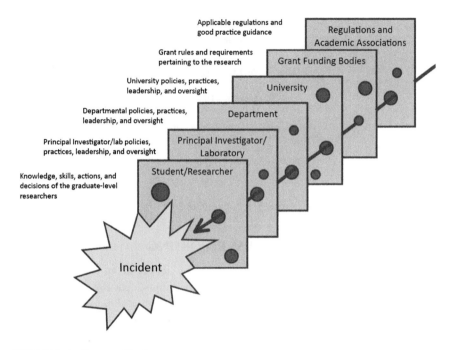

FIGURE 3.2 Layers of Defense.

Reason's "Swiss cheese model" illustrates the multiple layers of defense as slices of Swiss cheese and the holes within each slice represent opportunities where failure can occur. When a number of holes align, an incident results. The CSB used the "Swiss cheese model" to illustrate the multiple levels of failure that aligned to cause the laboratory explosion (Figure 3.2).

When an incident occurs, it is important to identify all of the gaps and failures that contributed to the incident and focus safety improvements at each of these levels in order to have a greater impact in preventing future incidents.

INCIDENT INVESTIGATION METHODOLOGIES

There are a number of incident investigation methodologies, and different methodologies may be chosen based on the severity of the incident. However, all incidents should be investigated to identify the direct, contributing, and root causes.

Direct causes, or immediate causes, are the aspects of the incident that immediately contributed to the injury or damage. An example of a direct cause of a laser eye injury is a specular reflection from a sensor card striking the eye during an alignment procedure.

Contributing causes are the symptoms, the unsafe acts or conditions that contributed to the incident. An example of a contributing cause of a laser eye injury is the failure to wear laser protective eyewear during alignment.

Root causes, or basic causes, are the underlying system or organizational related failures that lead to the unsafe acts or conditions. An example of a root cause of a laser eye injury is the failure of management to enforce the safety rule requiring laser protective eyewear, leading the laser user to believe that wearing eyewear is not really that important.

Many incident investigations stop at the contributing cause, the unsafe behavior or condition, without identifying the underlying causes. It is the identification and correction of the underlying causes that can lead to the most impactful and long-lasting changes.

As previously mentioned, the incident investigation method may be chosen based on the severity of the incident. The method may also be chosen based on the type of incident investigation that will be conducted, such as a one-on-one discussion or a more formal team-based investigation. A simple incident investigation method is the 5 Whys. The 5 Whys is a technique used to determine the root cause by repeating the question "Why?" at least five times. This method may be more suited for minor incidents or one-on-one discussions.

More serious or complex incidents may require a more sophisticated investigation technique, such as a causal tree. The causal tree method starts at the end of the event and works up the tree one level at a time, asking the following questions at each level:

1. What was the cause of this result?
2. What was directly *necessary* to cause the end result?
3. Are these factors (identified from question 2 above) *sufficient* to have caused the result?

Any root cause analysis method used must identify the critical events and conditions in the incident sequence and the management and organizational factors that allowed these events and conditions to occur.

TEAM-BASED INVESTIGATIONS

Incident investigations can benefit from a team-based approach, particularly when investigating more serious or complex incidents or incident trends. Team-based investigations are also valuable when the laser safety officer does not have direct laser experience.

The Center for Chemical Process Safety (CCPS) *Guidelines for Investigating Chemical Process Incidents* lists several advantages to using a team approach for investigating incidents, including:

1. Multiple technical perspectives assist in analyzing the findings. A formal analysis process must be used to reach conclusions, and individuals with diverse skills and perspectives best support this approach.
2. Diverse personal viewpoints enhance objectivity. A team is less likely to be subjective or biased in its conclusions, and the conclusions are more likely to be accepted.
3. Internal peer reviews can enhance quality. Team members with relevant knowledge of the analysis process are better prepared to review each other's work and provide constructive critique.
4. Additional resources are available. A formal investigation can involve a great deal of work that may exceed the capabilities of one person.

In a research setting, a team may be comprised of the laser safety officer, members of the laboratory experiencing the incident, peer researchers, and departmental safety personnel.

CASE STUDY OF A SUCCESSFUL TEAM-BASED INCIDENT INVESTIGATION PROCESS

One university changed its approach to investigating incidents after three minor incidents occurred in a laser lab within one year. The first two incidents involved small fires that caused minimal damage. These incidents were investigated using a single investigator from the environmental, health and safety (EHS) department who identified the direct and contributing causes. While the recommendations resulting from the investigation were implemented, the underlying causes were not identified or addressed, and another incident occurred. The third incident involved a small explosion in a glass vial that resulted in a minor injury to a graduate student. After the third incident, the environmental, health and safety department evaluated their incident investigation process, and determined that it was not leading to improvements in the safety culture and safety management in the lab. The EHS department decided to use a team-based approach to investigate the third incident, and reached out to the department chair for support and recommendations for peer researchers to serve as team members. The team was comprised of the laser safety officer, who worked in the EHS department, the principal investigator and two members of the affected lab, two peer principal investigators and two graduate students from similar labs, and the departmental safety representative. The chair charged the team, outlining the expectations for the team and asking for recommendations that could be applied across the department.

The peer researchers led the investigation process, with the laser safety officer facilitating the conversation. The advantages to having the peer researchers lead the process included their high level of technical knowledge and their understanding of the challenges of managing safety in a university laboratory with limited resources and high administrative burdens. The lab in which this incident occurred did not have a lab manager. Instead, there was a graduate student informally identified as a safety liaison. The team discussed the feasibility of hiring a lab manager, but the limited financial resources did not allow for this. The team discussed establishing a formal safety leader role for a senior graduate student in the lab, with clear responsibilities, training, and a process to mentor a successor. The team recommendations included improvements to the experimental procedure and controls and the establishment of the safety leader role. These recommendations were presented to the chair, who was able to implement them at the departmental level. Feedback from the principal investigator from the lab involved in the incidents was positive, and he appreciated the collegial nature of the investigation and the level of expertise and understanding of the team members.

This process was used again when an incident occurred in a different laser lab that involved an exposure to a specular reflection from a Class 4 laser. The graduate student briefly removed his laser protective eyewear during an alignment procedure.

During the team-based investigation, the principal investigator of that lab discussed the lack of consistent enforcement of the laser eyewear policy. The fact that he self-identified this problem and discussed this with his lab group was impactful and led to an improved safety culture in his lab, and a lesson learned that could be shared with other labs at the university.

CONCLUSION

While the focus on safety culture continues to increase, not all organizations have a strong, generative culture of safety or the systems in place to support one. A strong safety culture requires a commitment from the top to make safety a priority throughout the organization. While the laser safety officer is often not in the position to affect the safety culture of the entire organization, there are opportunities to support and develop a positive safety culture within the laser safety program. The laser safety officer can establish relationships of trust and respect with the laser users. When this relationship is in place, lab visits can become opportunities to learn about near misses that may have occurred in the lab, and laser users will be more likely to reach out to the laser safety officer to get assistance in correcting unsafe conditions. Some laser safety officers are laser users or have previous experience as laser users, but some come from a health and safety background and have never operated a laser. In those instances, the laser safety officer can benefit from the relationships with the laser uses and employ their technical knowledge when needed.

The goal is always to prevent incidents from occurring, but incidents do still occur. It is important to understand how safety culture and incident reporting and investigation are intertwined, and the importance of using incidents that do occur as an opportunity to learn and improve.

4 Accident Preparation

Ken Barat

CONTENTS

Let's start out like this. It has been a long, long, long week. It is Friday, 4:30 p.m., and you have been looking at the clock for the last 2 hours. Just 30 more minutes and you will be facing the commute home to hopefully a stress-free weekend. Then it happens, the phone rings: *Is this the LSO? I think I we have a laser eye injury?*

After some unprintable words pass through your mind or lips, how do you respond?

Option A

I am on my way home, I will see you on Monday, call your supervisor or have your supervisor call me (if he can find me, I am out of here).

Well, you have just told whoever called that you really do not care what happened and it can wait. Your next concern is will you still have a job on Monday?

Option B

Who is this? Where are you located? Is anyone with you? Stay calm, we can take care of this.

Your preplanning for an incident now goes into effect.

So, which response would you want to hear or are prepared to give?

BEING PREPARED

Traditional laser safety training talks about laser accidents, common causes, who they happen to, but rarely talks about the real logistics of the event. So, let's review some of the real-world questions one needs to be prepared to deal with, starting with:

- Do staff know who to call if an injury is suspected?
- Is the Laser Safety Officer prepared to respond?
- Has medical assistance been selected?
- Do users know how to respond?
- Is there an investigation strategy?
- What about lessons learned?
- Are regulatory actions required?

Do Staff Know Who to Call if an Injury is Suspected?

This seems like a simple question, but in fact, it is very important. Where and how has this information been relayed to staff. There are a few traditional ways and less "traditional or modern means."

The most common approach is an Emergency Response Poster. This can be a poster size sheet or a flip chart with instructions for different crises. Fire, Chemical spill, Radiation Spill, Bomb Threat, Electrical Shock, and the list goes on.

On these types of flip charts, I have rarely seen laser injury get its own chart or title page. More commonly, it is mixed in with some other injury type; Personnel Injury as an example. Like any poster or sign, these become invisible after a while and sometimes get covered up with other announcements.

The typical single page poster (of expanded size) is usually a list of personnel responsible for different safety disciplines: Electrical, Chemical, Radiation, Laser, and so on. The error here is unless the position is assigned a permanent phone number, people come and go and the poster does not get updated or taken down. So, you could have a poster listing Edison as an electrical contact, but as we know, Thomas is long gone.

Another approach is to have a single emergency number that all are supposed to know and remember, such as 911. Of course, depending on your phone system, only 4-digit extensions are accepted and your 911 might be 5911.

Another approach is to have the information on a bookmarked web page; of course, one needs a tablet, smartphone, or terminal to reach it. The web approach does allow for rapid updating of people and extension.

Some institutions have included what to do if a laser injury is suspected as part of laser standard operating procedures. The instructions could be reviewed as part of the SOP renewal cycle. Of course, you are hoping people can find the SOP once it is signed off.

Standard Operating Procedures (SOP): What Should Be Included?

Emergency Response

Authorized laser users will be familiar with the Building Emergency Plan, location of emergency equipment, and emergency procedures for fires, earthquakes, and evacuations. Emergency shut-off for lasers is done at the electrical panel circuit breakers or by electrical shut off switches.

Suspected Laser Injury

Accidental laser beam exposure is a serious event. In the case of suspected laser injury, operations will be ceased and the laser set-up will remain unchanged to allow for analysis of the cause of the accident. Exposed employees will be transported to Health Services (Bldg. xxx) for evaluation. Call A-911. Notify any staff in the area, laser use supervisor, and LSO. Key item is to keep the individual calm.

Finally, put the information on a downloadable phone application. This way, staff might have the information with them, no matter where they are in the facility.

Regardless of what approach is decided upon, staff need to know how and what to do if an injury is suspected.

So, what should we do?

I Suspect an Injury to Myself

Well this only applies to an eye injury. Skin burns are rather yes or no. A laser skin burn requires standard first aid attention.

Now we come to the suspected eye injury.

Ultra Violet Wavelengths

These wavelengths can cause cornea and lens injury. The nature of UV cornea injury is that there is no immediate sensation of injury, so the pain response does not manifest itself for on average 10 hours. So, many times, the person is away from the work environment when they feel pain. Which generally facilities a trip to the ER.

Visible and Near Infrared Wavelengths

Signs of possible exposure go from the extreme to mild. There is the nanosecond pulse popping sound from inside the eye (cavitation bubble bursting), spots floating in the field of view, to field of view becoming red from bleeding into the eye, to dark spots in the field of view, to a vague or blurry spot outside the field of view.

IS THE LASER SAFETY OFFICER PREPARED TO RESPOND?

Seems like a straightforward question. In the real world, it is not so easy. The answer relates to have you thought about what you need to respond and one's level of confidence (which can be too low or too high). Now that you have been contacted:

- How soon can you get to the site?
- What do you want to bring with you?
- Are there others who should come with you?
- Who should you notify of the suspected incident?
- What do you hope to accomplish?

HAS MEDICAL ASSISTANCE BEEN SELECTED?

Depending on the size and structure of the institution, it may have its own on-site medical unit. Their function is usually first aid and on-site physicals. They may do eye examinations. Are they prepared to deal with a suspected laser eye injury? If the answer is no, then they need have a resource to direct the individual to or they need education.

One may have to contact the local State Ophthalmology Society to find the correct people to go to.

In most metropolitan cities, there is at least one hospital designated as a "Trauma Center," that may be your choice. Once again, the LSO should determine which medical center is the right place to send your users.

DO USERS KNOW HOW TO RESPOND?

What I mean by this is do those around the individual know how to respond. The key is to keep the person calm and prevent them from going into shock. One can die from shock, but one cannot directly die from an eye injury.

IS THERE AN INVESTIGATION STRATEGY?

Once a suspected injury is reported and the individual taken care of, it is time to think about finding out what happened. The bottom line is to find out: What caused the incident? Was it a pilot error, or is there a flaw that might happen again?

WHAT ABOUT LESSONS LEARNED?

How do we get the word out to alert people of what happened and how they can avoid becoming a victim? Remember lessons learned is not hunting for a head to remove; rather, it's what can be done to prevent the event from happening again.

ARE REGULATORY ACTIONS REQUIRED?

In the regulatory chapter (Chapter 12), we discuss timelines requested by regulatory agencies to notify them of laser injuries. If your State has a laser regulatory program, there always seems to be a time period for reporting laser injuries: from 2–24 hours is common. If there is no State program, it might just be an OSHA reportable item.

OSHA Reportable Event

1904.7(a) Basic requirement. You must consider an injury or illness to meet the general recording criteria, and therefore to be recordable, if it results in any of the following: death, days away from work, restricted work or transfer to another job, medical treatment beyond first aid, or loss of consciousness.

1904.7(b)(1)
How do I decide if a case meets one or more of the general recording criteria?

A work-related injury or illness must be recorded if it results in one or more of the following:

1904.7(b)(1)(i)
Death. See § 1904.7(b)(2).
1904.7(b)(1)(ii)
Days away from work. See § 1904.7(b)(3).
1904.7(b)(1)(iii)
Restricted work or transfer to another job. See § 1904.7(b)(4).
1904.7(b)(1)(iv)
Medical treatment beyond first aid. See § 1904.7(b)(5).
1904.7(b)(1)(v)
Loss of consciousness. See § 1904.7(b)(6).
1904.7(b)(1)(vi)
A significant injury or illness diagnosed by a physician or other licensed health care professional. See § 1904.7(b)(7).

A FEW WORDS ON LASER MEDICAL EXAMINATIONS

While we are in a hurry to get the individual examined, some laser defects may take a day to appear, so a return visit within 24 hours might be necessary or advisable.

IS THERE A VALUE TO PERIODIC EYE EXAMS?

People have asked: "I see laser light, flash lamps, pulses, and so on. Does this have an effect on my vision?" Excluding an acute exposure, to date, no chronic health problems have been linked to working with lasers. Also, most uses of lasers do not result in chronic exposure of employees even to low levels of radiation. Many eye examinations have been performed over the years, and no indication of any detectable biological change was noted.

WHAT WERE THE STANDARD LASER EYE EXAM ELEMENTS?

The following items made up the standard "Baseline" laser eye examination. One point many times overlooked was the laser standard only required an ocular fundus exam, if any of the first four exam elements was abnormal. The fundus exam was always considered an addition test. But many institutions included it as a routine member of the exam. Therefore, unless a digital camera was used, it required dilation of the eyes and exposure to the bright light of a slit lamp. This is the genius of horror stories associated with the baseline laser eye exam.

The Baseline Eye Exam Requirements—No Longer Required by Laser Standards

1. Ocular History • Inquiry should also be made into the general health status with a special emphasis upon systemic diseases which might produce ocular problems
2. Visual Acuity (20/20 with corrections, if necessary)

3. Macular Function
4. Color Vision examination

Note: Only if any of the above items was abnormal should Step 5 be performed.

5. Examination of the Ocular Fundus with an Ophthalmoscope or Appropriate Fundus Lens at a Slit Lamp

The following items may also be observed and documented:

- Presence or absence of opacities in the media
- Sharpness of outline of the optic disc
- Color of the optic disk
- Depth of the physiological cup, if present
- Ratio of the size of the retinal veins to that of the retinal arteries
- Presence or absence of a well-defined macula
- Presence or absence of a foveal reflex

Any retinal pathology that can be seen with an ophthalmoscope (hyperpigmentation, depigmentation, retinal degeneration, exudate, as well as any induced pathology associated with changes in macular function).

5 Lessons Learned Examples

Ken Barat

CONTENTS

While the value of lessons learned is well established, finding examples can be a challenge. This chapter is composed of 10 lessons learned that you can reprint for your use. The examples cover laser hits in the eye, but also items such as ergonomic concerns, eyewear flaws, seismic safety, and others.

A lesson learned program should not just consist of sad or scary stories, but involve a mixture of good practices and informational items for staff. Fortunately, or rather unfortunately, the items included do not get stale or outdated. I have tried to follow a single template for each example. This template or layout can easily be adopted to whatever format works best for you. Here are the 10 examples.

EXAMPLE 1: TI:SAPPHIRE STRIKES AGAIN

At a Glance

A post-doctoral researcher (PR) was aligning the beam of an 800 nm, femto-second repetitively pulsed, Class 4, Ti:sapphire laser when he lowered his eyewear in order to better locate the laser beam. The PR turned and observed a subtle flash of light strike his right eye. Later in the evening, he noticed a blind spot in his vision.

Details

This activity started a month earlier with initial planning and set-up. The laser's 2.5 watt beam was split sending ~50 milliwatts down a time-delayed leg to three retroreflectors. The mounting fixture was too small to fit three "shielded" retro-reflectors, so the PR installed a legacy "unshielded" retro-reflector in the middle position. The worker forgot to check for stray beams around the retroreflectors.

On the morning of the accident, the worker and the laser system supervisor (LSS) aligned the system to the cryostat. Prior to beginning the alignment process, they did not recheck previously installed optics for stray beams.

In the afternoon as the PR was making fine adjustments, aligning the beam through a super continuum-generating crystal, he chose not to use the IR viewer or viewer card used in the initial alignments, as he assumed those tools would not provide him with the visual acuity he needed for the precision alignment task being performed.

A neutral density filter had been installed in close proximity to the crystal. This reduced the beam power entering the crystal during alignment, but did not reduce the power of the rest of the system upstream of the crystal.

During the alignment task, the PR turned his head toward the higher power upstream beam path and was struck in the eye by a stray beam reflected off of the "unshielded" retro-reflector.

The Lesson/Word of Caution

Near infrared beams have a great potential for injury. Remote viewing, beam containment, and checking for stray reflections are your best protection. Eyewear only helps if you use it.

EXAMPLE 2: LASER HAZARD WITH RESPECT TO HEIGHT

At a Glance

It is common for laser users to be told to not position the laser beam path at "eye" height, sitting or standing. I think most laser users feel this is good advice but hard to adhere to. For the fact is, there is no ideal eye height.

The best one can do is to see that any stray reflections from the set-up are blocked so they do not escape to the defined laser set-up space. I have seen many a workstation that is in direct line with the optical table and optics for the sitting researchers, with and without protection for the staff, such as a perimeter guard.

Figure 5.1 proves again that nothing in our lives is as simple as we might like it to be.

As you can see, a percentage of the beam goes over the beam block, which can present a different level of hazard depending on one's height. Which is one of the reasons why I am very concerned about the height of some of the beam blocks and perimeter walls I have observed being used, see Figure 5.2a,b.

EXAMPLE 3: THE 5 ITEMS THAT GO INTO EYEWEAR SELECTION

At a Glance

Laser protective eyewear is one of the most common and relied upon laser safety devices. It should be said that with proper beam containment, eyewear may not be

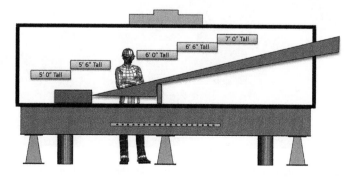

FIGURE 5.1 Beam height can change.

FIGURE 5.2 (a) Table guards, (b) Guard barely taller than optical mount, (c) Razor style beam block.

needed. The use of remote viewing with CCD cameras can also reduce the need for laser protective eyewear. When it is needed, how does one select the best eyewear to provide protection and visibility?

Key items in the selection of laser protection eyewear:

- Wavelength coverage
- Optical Density (OD)
- Visual light transmission (VLT)
- Fit
- Visual Acuity

Wavelength Coverage

First step is to find a filter that covers the wavelength range one will be exposed to. With ultrafast lasers, the astute user will consider the bandwidth of the pulse. Many filters designed for ultrafast lasers offer a very narrow coverage bandwidth, this is to increase VLT. Which is why reflective filters have become more popular. Be aware, these filters can reflect sufficient energy to be an eye hazard to others. Information on coverage can be obtained from vendors.

Optical Density

OD is the attenuation factor of the eyewear. The OD required is calculated from output specifications which yield the irradiance the beam could deposit on the eye (depending on wavelength, the retina, or lens). It is a unit less number in base 10. So, an OD calculation that says 3.4 OD is needed requires an OD 4 filter, not OD 3. Picking a filter with an excessive OD higher than needed, at times, will cause a loss in VLT and or acuity. The calculated attenuation of the filter is to reduce any direct beam exposure to a safe exposure level, termed Maximum Permissible Exposure (MPE). MPE is based on biological testing data.

Visual Light Transmission

This is given as a percentage of visible light that is transmitted through the filter. Under 20% is not desirable. Overall, the higher the better, with exceptions (see Visual Acuity). As a general rule, glass eyewear or reflective eyewear will yield higher VLTs than plastic filters. But weigh more on one's face.

Visual Acuity

While it seems counterintuitive, by comparing two filters side by side, the one with the higher VLT may not always yield the best acuity—ability to see. Just like how blue blocker sunglasses, which give good vision, have a VLT lower than 20%. One needs to look at the vendor website or get a sample pair to evaluate their acuity. Efforts are underway to have a figure for VA listed for filters.

Fit

The frame must fit your face and feel comfortable when worn. If not, it will not be used or cause people to lift it up or pull them down. Great strides have been made in frame design for laser users. It is rare that one cannot find a comfortable pair to wear. Please note glass filters, which have a higher VLT, are heavier on the face than plastic eyewear and may require a strap to keep them from slipping on one's face. Prescription glasses wearers will either need over-the-frame style laser frames or use a frame that has a prescription insert option.

Three Special Considerations

Eyewear using dielectric coatings need care in handling—particularly when not being used. Do not lay down lens on surfaces. The coatings can be scratched off. Even if they say they have a protective coating, care is still required. Defects may be too small to observe.

Due to the slow relaxation time of absorptive filters when compared to ultrafast pulses, the stated OD may be less when struck by a chain of ultrafast pulses—Pico and femtosecond are the pulses I am referencing here—making beam containment and remote viewing even more critical to user safety.

Not every filter comes in every frame style; this relates to the curvature of the frame and associated difficulty of getting a uniform OD across the curve surface.

LSO Assistance

A user can always confer with the LSO in the selection process.

EXAMPLE 4: INCOMPLETE WAVELENGTH COVERAGE

At a Glance

On January 12, 2016, a Sandia National Laboratory (SNL) laser operator experienced a flash in his eye when aligning a laser beam from an 800-nanometer wavelength laser source. The operation takes an 800-nanometer wavelength beam and converts it to a 400-nanometer wavelength beam. The beam exceeded the height of the beam block set in place, and the eyewear did not cover 400 nm. After experiencing the flash, the operator stopped the beam alignment operation, secured the equipment, and reported the incident to management.

Details

The laser operator was wearing protective personal equipment that was appropriate for 800-nanometer wavelength but not for the 400-nanometer beam. Protective eyewear that covers both wavelengths was available and present in the lab. The presence of a vertical beam had been discussed during a pre-job briefing on how to set-up the experiment. There was no discussion on checking the location or height of the beam, it was assumed that would be a natural action (skill of the craft) by the user. The pre-job briefing was held by the laser operator, Principal Investigator, and the laboratory owner on the day before the experiment.

The Lesson/Word of Caution

One must not only know the wavelengths being used but confirm any eyewear used is correct. You would not use the eyewear if the lens were cracked. Well, if it does not cover the wavelengths in use at the appropriate optical density, it might as well have a gaping hole in it.

Medical Outcome

SNL Medical did not find any eye injury or abnormalities to the laser operator.
   Special note: To avoid a similar incident from occurring:

- Always use the appropriate laser protective eyewear that has full protection for ALL wavelengths in use.
- Utilize adequate beam blocks: appropriate size and quality to withstand the beam strength and conditions.

EXAMPLE 5: FLUORESCENCE FROM LASER EYEWEAR

At a Glance

A researcher wearing eyewear noticed blight flashes on the surface of the eyewear.

Details

On July 18, 2016, a worker at a DOE Lab was aligning a pulsed green laser (527 nm) to a semiconductor wafer when a reflection was directed off the face of the wafer toward the worker's laser protective eyewear (LPE). The worker described the laser beam interaction with the LPE as bright flashes of light across the upper part of the eyewear. The worker was concerned because what he saw (bright flashes) was similar to what is reported by those involved in laser eye injuries. Management was notified and the worker was directed to get an eye examination. No damage was found.

   Believing that the LPE may be defective, tests were performed on the eyewear. Good news: it was found to be working as designed. The eyewear did brightly fluoresce when struck by a green laser beam (see Figure 5.3). Other eyewear in the lab was tested, and only the YAG/KTP filter was found to produce the fluorescing phenomenon when struck. The rest produced only a dull spot on the inside of the eyewear during the same test.

   What about other manufacturers? Further testing found other filters by different manufacturers resulted in the same fluorescence. Therefore, be aware when using filters used to block visible laser beams, there will be some reemission of light if stuck by a laser beam.

   Key item: Anytime one notices any effect on their eyewear, from a flash, flashes, smoke, blurring to composition of the eyewear, then step to the side and examine the

FIGURE 5.3   Filter fluorescence.

eyewear. It is best to first leave the laser area or block the beam, before removing one's eyewear.

Example 6: Falling Pump Beam Tube Leads to Reflection in the Eye

At a Glance

A researcher was diagnosing a power loss from a Spectra Physics 3900 Ti:sapphire (oscillator) that is pumped by a Millennia eV DPSS laser. During this procedure, they noticed some leakage in the form of a green diffuse light being emitted from the end of the metal tube connecting the DPSS pump laser to the Ti:sapphire laser. The researcher attempted to readjust the beam tube to eliminate the leakage to improve the beam tube alignment. As he touched the beam tube, it became dislodged from its mounting, striking the 532 nm laser beam as it fell, causing the laser beam to reflect for a brief moment in many directions until the tube fell completely to the table (Figure 5.4).

Details

During the moment of reflection, both researchers saw a strong green light coming from the tube. Both researchers were wearing laser eye protection for the 850 nm wavelength Ti:sapphire laser, but not for the 532 nm wavelength, which is normally enclosed by the beam tube. Both researchers received eye exams, and fortunately, no eye damage was found.

   What corrective actions were taken? The researchers will replace the two-piece beam tube with one that is anodized black and spring loaded.

The Lesson

On a routine basis, laser users should check that beam stops, tubes, barriers, and so on, are in place and secured before turning on the laser system. In addition, make sure you are wearing the proper eyewear for the job. While they did not plan on the tube dropping and exposing them to green light, good practice would have had them wearing 523 nm eyewear as a precaution.

FIGURE 5.4 Beam tube failure.

EXAMPLE 7: ULTRAFAST PULSES AND LASER EYEWEAR

At a Glance

A study by the U.S. Army Public Health Command demonstrated that the unique properties of ultrashort laser pulses have an effect on laser protective eyewear filters. This effect can lower the optical density of the filters.

Details

Ultrashort pulses can cause shockwaves and bubble formation in the tissues of the eye, which lead to tissue damage.

An injury can occur from an intrabeam exposure or from viewing a reflection of the beam. Most documented injuries from ultrashort laser pulses have occurred when personnel view a reflection of the beam while eyewear has been removed or one is looking under/over the eyewear. Exposure to just a single pulse, occurring on the order of a trillion times shorter than the blink of an eye, is all that is needed to cause damage.

The unique properties of ultrashort laser pulses make it difficult to protect personnel from them. The lack of protection, in this case from a lower optical density (OD) than that is specified by a manufacturer for such short pulse durations, can be contributed to several factors:

- High peak power: Saturable absorption in the filter material may occur, which reduces the OD.
- Absorptive material, does not have a quick enough relaxation time.
- Broadband emission: Supercontinuum generation converts a narrow wavelength band to a wide wavelength band which might require a broadband filter.

So How Do I Protect Myself?

Remote viewing is the second-best option. Many common laser wavelengths, including those produced by the Ti:sapphire laser commonly used to create ultrashort pulses, can be viewed with inexpensive webcams or digital cameras. Another option is, when feasible, alignments should be made with the laser operating at a reduced energy or a longer pulse width.

The best option for protection during operation would be to enclose the beam path of the laser when emitting ultrashort pulses. This has the added benefit of protecting the optical components from dust.

What About M Rated Laser Protective Eyewear?

M rated eyewear is narrow notch eyewear made to withstand ultrashort pulses. The problem is the few filters made this way are for very narrow wavelength bands.

So, Should I Wear My Existing Eyewear?

The answer is YES, but awareness that possible reduction in OD is possible from a direct hit, should make one even more eager to follow good practices.

Material for Example 7 comes from U.S. Army Public Health Command, *FACT SHEET 25-026-0614.*

EXAMPLE 8: IR SENSOR CARD

At a Glance

Just because nothing has happened at your set-up does not mean vigilance can be let down. People come and go, and all need to understand and respect the potential for an incident to occur. Remember a lack of incidents does not always mean that good technique is being followed.

Phosphorescence

Even though infrared light has a too low photon energy for directly producing visible fluorescence (except with upconversion processes), it can trigger the emission of phosphorescent light.

The probably most common sensor cards for the near infrared wavelength region (for example, used for working with YAG lasers) contain some kind of phosphor. Before use, the phosphor must be "charged" by illumination with visible light—sunlight or artificial light. The material then generates a low level of phosphorescence (afterglow), which can hardly be recognized. However, when hit by infrared light, the material releases the energy stored during the charging process much more quickly, and the illuminated spot can be seen (often in orange color) with the naked eye. This visual impression lasts only until the stored energy has been exhausted; one then needs to recharge the card or just move the laser spot to another location which has not yet been used.

Depending on the used phosphor, the wavelength region with good sensitivity can be narrower or broader. The wavelength range specified by the manufacturer often includes regions where the sensitivity is actually very low, while quoted numbers for the sensitivity often apply to wavelengths with maximum sensitivity. Under ideal conditions (optimal wavelength region, short-term use after charging the card with bright visible light), some cards allow one to detect infrared intensities of the order of a few $\mu W/cm^2$, whereas in other cases, one requires multiple mW/cm^2.

Upconversion Fluorescence

There are rare-earth-doped materials which can upconvert infrared radiation to the visible region. Such materials do not need any charging before use. However, the upconversion process requires higher optical intensities, that is, the sensitivity of such cards is substantially lower. (It can be better with pulsed light.)

Direct Generation of Fluorescence

Laser viewing cards for ultraviolet light can directly produce visible fluorescence light and do not need any charging before use. Because such processes can be very efficient, the achieved sensitivity is usually quite high. In some cases, there is also substantial level of phosphorescence, which may last for several minutes.

Thermochromic Materials

The thermochromic operation principle can work in a very wide wavelength region.

Particularly for use in the mid-infrared spectral region, thermochromic detection cards are available. These contain a material (for example, a liquid-crystal film or a material containing certain pigments) which changes its color when it is heated to an elevated temperature. Depending on the type of material, the color may change, for example, from green to brown or black, or from dark to light.

This operation principle means that the laser intensity must be high enough (for example, above 100 mW/cm^2) to raise the local temperature by some tens of Kelvins. On the other hand, the laser intensity should not be far higher, because otherwise, the material could be destroyed. Therefore, the usable dynamic range is fairly small—often with a factor less than 10 between the damage threshold and the minimum detectable intensity.

The visible spot remains until the material has cooled down again. That process may be accelerated by pressing the card against some cool metal part, for example. If the ambient temperature is too high, it may not be possible to use such a card.

An advantage of the thermochromic operation principle is that it works in a very wide wavelength range, because only sufficiently high absorption of the laser light is required to achieve the temperature rise. For shorter wavelengths, however, other cards (for example, based on phosphorescence) are normally preferred due to their much higher sensitivity and larger dynamic range.

Practical Aspects

Laser viewing cards are often used in reflective mode—one observes the generated visible light on the same side where the laser beam hits the card. Some cards, however, can also be used in transmission—one observes on the opposite side. The sensitivity is usually lower for use in transmission.

In some cases, a sensor card contains different photosensitive areas, for example, different wavelength regions or different sensitivities.

Sensor cards often also work for visible light, but in this case, there is normally no advantage over simply using a piece of paper, for example, as a scattering surface.

Putting reflective tools into laser beams can be dangerous!

Note that part of the incident laser power can be reflected by the top surface of a sensor card, and this can cause a safety hazard when working without laser goggles, because laser light may be directed toward an eye.

Most detector cards are just held with the hand or possibly fixed in some clamp, but there are also cards specially made for mounting in standard optics posts. Mounting a sensor card at a fixed position is of course less appropriate for cards which need recharging; one would rather move such a card around in order to use different spots of its active area.

Related Devices

IR Sensor Cards by LUMITEK International work on a principle known as "Electron Trapping," where phosphor-based compounds are employed to absorb and "trap" incoming light energy from a short wavelength, and release that stored light in the form of visible light upon stimulation from a longer IR wavelength. The visible result is a localized glow, which is relative in intensity to the amount of stored light and IR power levels exciting the active area.

Based on a special durable ceramic-like material, the detectors combine high sensitivity with high damage threshold. In contrast to other conventional infrared-to-visible converters, the IR-VIS series from ALPHALAS do not need any activation with daylight or UV light and can therefore be used in darkened rooms. In addition, the irradiated area does not bleach. If the surface of the active area has been damaged, the upper layer can be simply removed using a fine file or sand paper. The new fresh surface is as good as the original one.

The standard -S and -D models cover two IR spectral ranges from 800 nm to 1100 nm and from 1460 nm to 1600 nm, resulting in green and orange emission correspondingly. The diameter of the active area is 15 mm or 40 mm. The -S models are single-sided and the -D model is double-sided.

The special -QM model is for Q-switched and modelocked lasers only. It is based on an efficient second-harmonic generation process and covers the wide spectral range from 800 nm to 1400 nm. It is also double-sided and has a diameter of 30 mm.

The new IR-conversion screen is a unique card. Starting below 800 nm up to 1600 nm, it converts invisible light into visible. Its emission is red (except at incidence of 950–980 nm where it emits green). The emission is based on "photon upconversion" effects. The LDT-008 is active, it does not need to be excited with UV-light. Converting from 1 mW up to 5 W/cm^2, 1064 nm (even higher for a short period), it is a very versatile tool for everybody working with lasers in the range of 800–1600 nm. The active area is 54 × 42 mm and it is transmissive to allow the signal to be seen from the rear.

EXAMPLE 9: LAB ERGONOMICS

At a Glance: Lab & Experimental Layout Planning Will Prevent Back Injuries and Other Ergonomic Problems

The three basic ergonomic factors in the lab are posture/position, repetition/duration, and force. What I would like to talk about in this laser safety note is position and posture. Remember your Mom saying sit up straight?

The most common issue in the laser lab relates to reaching for optics, equipment, and the strain it can place on one's body. Since one never has sufficient room in their laser work space, being able to reach all optics, and so on, can become a challenge. Many times, one cannot arrange the lab so access to the optical table is possible from all sides. In some cases, pieces of equipment become the instrument that makes accessibility difficult.

Depending on how often one must reach for these optics, it might not seem like a concern. Please keep in mind one minor torqueing of your back or body may be the start of a lifetime of pain and discomfort.

What are some solutions? Each of these solutions below deserve some consideration and will not be the answer to all set-ups, but are worth considering.

1. Motorized mounts: Such mounts are controlled off the table and make reach in unnecessary and should be partnered with remote viewing cameras
2. Platforms to stand on giving greater reach in: These come in different heights and varies, numerous commercial platforms are available

3. Vertical bread boards on table: At times, can bring optics closer to the reach of users
4. Set-up so one can reach in from all sides, or at least three: This is just good planning, take advantage of building codes to get extra room
5. Raising some items on shelves: While this means elevating some beams, the risk might be balanced off by increased accessibility
6. Optical table modifications (some optical tables can be configured with a hole in the middle, like model railroad tables for access, and still maintain optical table properties): While I have rarely seen anyone use this option, this might be worth considering in initial lab set-up
7. Classical ergonomics: Standard ergonomics for the lab, height and position of monitors, where keyboards are placed, the type of mouse used, and so on. These items are the same as routine office ergonomic safety, but are rarely considered in lab set-ups. The usual reason given is about how little time a person spends working with them. To me, that is like saying I do not use my seat belt on short trips.

   The items in solution 7 need to be looked at by someone familiar with ergonomics. The typical rules are:
   - Monitors should be directly in front of you with the upper edge at eye height or slightly below; needs to be adjustable.
   - Use a document holder for hardcopies and keep in front of you, between monitor and keyboard.
   - Keep the keyboard and mouse in front of you and as close as practical to prevent over reaching.
   - Keep wrists as straight as possible.
8. Trip hazards: Trip hazards need to be addressed. This can be done by "bridges" designed to contain wires and hoses.

The Occupational Safety and Health Administration provides a review and recommendations through its eTool on ergonomics.

Note: This list of solutions was inspired by an article by Vince McLeod in *Lab Manager* 12, no. 4 (May 2017).

Example 10: Three Overlooked Non-Beam Hazards in the Laser Lab

At a Glance: Seismic Safety, Optical Table
Grounding, and Fiber-Optic Safety

Seismic safety is often overlooked due to the complexity of the issue and cost. Optical table grounding is also overlooked for users thinking the use of grounded commercial equipment takes away the need. Lastly, fiber-optic safety many times goes unaddressed for a lack of understanding of the risk.

Details: Non-Beam Hazards

While Class 3B and Class 4 lasers have the potential to be a serious eye hazard, they are not the only hazards in our laser use areas. There are several non-beam hazards common to many laser use areas. Sometimes, they are referred to as "Associated Hazards," for those who prefer a continental term. Three overlooked or underestimated non-beam hazards include earthquake restraints, optical table grounding, and fiber-optic safety.

Earthquake Restraints

While the building structure is rated for a 7–9.5 shaker, most injuries and concerns are over items inside the laser use areas. This ranges from items falling from height above 5 feet to movement of heavy equipment that could block one's exit from the laser use room (Figure 5.5).

(a)

(b)

FIGURE 5.5 (a) Earthquake aftermath lab setting, (b) Earthquake aftermath office setting.

Some solutions and items to be aware of are:

1. Chaining equipment racks to wall or floor to prevent them from dancing in front of your way out.
2. Bracing of optical tables for the same reasons as above (use of proper earthquake restraints legs).
3. Locking and securing items on shelves above you. Possible solutions include the use of heavy duty Velcro to keep equipment in place, which might be better than other options such as lips on shelves or chains across shelves.
4. Another good practice is to place fluorescent arrows on the floor indicating the way out or on the edges of doorway curtains.

Note: Items 1 & 2 are facility projects.

Grounding, is an Element of Electrical Safety

All optical tables need to be electrical grounded, and this could be a relatively simple thing to do. The modern optical table top is isolated from the rest of the table. This is to prevent anything that might get spilled on the table from getting inside the table.

Fiber-Optic Safety

While not every laser set-up has fiber concerns, when one cuts or trims fibers, a well-documented safety concern exists—being stabbed by fiber shards. Standard fiber safety involves using a "sharps container" for fiber cuttings, working on a dark background to make it easier to see them, no food around fiber work, wearing a lab coat (so fiber scraps, which are known to dance around better than jumping beans, don't get on one's clothing and thus be transported home). As long as I am mentioning fiber scraps, careful use of single-edge razor blades and their proper storage is more than a good idea.

Figures 5.6a–e drive home my points.

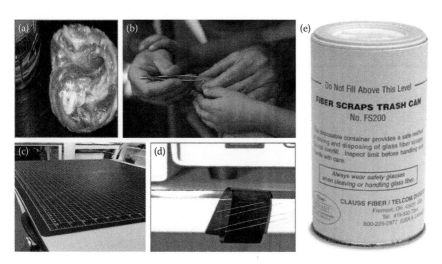

FIGURE 5.6   (a) Fiber contamination on food, (b) Fiber in finger (c) Black work surface, (d) Tape holding fiber scraps, (e) Fiber scrap can.

ADDITIONAL REFERENCES

Lyon, T.L., and Marshall, W.J. Nonlinear Properties of Optical Filters—Implications for Laser Safety. *Health Phys.* 51, no. 1 (1986):95–96.

Mclin, L. A Case Study of a Bilateral Femtosecond Laser Injury, *Proceeding of the International Laser Safety Conference 2013*, Laser Institute of America, paper #904.

Rockwell, B., Thomas, R., and Vogel A. Ultrashort Laser Pulse Retinal Damage Mechanisms and their Impact on Thresholds, *Medical Laser Application*, 25 (2010): 84–92.

Stolarski, J., Hayes K.L., Thomas R.J., Noojin, G.D., Stolarski, D.J., and Rockwell, B.A. Laser Eye Protection Bleaching with Femtosecond Exposure. *Proc. SPIE 4953, Laser and Noncoherent Light Ocular Effects: Epidemiology, Prevention, and Treatment III* 4953 (2003):177–184.

6 The Near Miss

Ken Barat

CONTENTS

The concept of reporting a near miss incident may seem strange to many laser users. It's almost of a "no harm, no foul" mentality. The majority of laser-related near misses are not so much the beam whizzing past someone's ear but rather procedural violation. A diode/fiber found on when expected to be off. A cover removed and the protective housing interlock found bypassed. The importance of near miss reporting is the chance to work on improving procedures as well as improving one's culture.

I would like to relate a near miss incident where I felt the reaction of the parties involved showed a strong positive safety culture, which could have easily gone another way. Which is what I would expect would have been the course of action at many facilities.

NEAR MISS INCIDENT HANDLED THE CORRECT WAY

The lab has several laser users and all have received instructional laser safety training. All have received some level of on-the-job training. The laser lab is equipped with a laser access interlock system. The status of the laser system is triggered by a shutter and warning light. When the shutter is down blocking the beam from exiting the laser box, the sign outside reads Class 1 condition (SAFE). When the shutter is in the open position, this allows a laser beam to exit the laser box and the warning light reads Class 4 Condition-Danger.

Day 1 of the incident: a user is working on an optical set-up outside the laser box, but in an enclosure that has perimeter guards to contain most stray reflections. In order to do the optic work, the users removed the shutter (see Figure 6.1), to give him more space to work. At the completion of his work, his intention was to replace the shutter, but this was not done. Meaning the warning light was in the closed position sending a signal to the warning light that all laser radiation was contained.

Day 2: a new user enters the room, no eyewear is worn since the door message says all is safe so laser protective eyewear is not required. He reaches into the area where the new optics have been placed. At that time, he observes a fluorescence on the sleeve of his shirt. Then he noticed the shutter is not in place. He places a beam block in front of the laser box aperture.

Our Day-2 user now has two options:

OPTION 1

Like a hunter, track down the person who left this trap for him and take his revenge. Once that is completed, put the shutter back and get on with work, as well as getting promises this will never happen again.

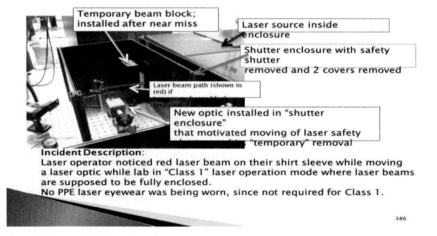

SLAC "Near Miss" Incident

Temporary beam block; installed after near miss

Laser source inside enclosure

Shutter enclosure with safety shutter removed and 2 covers removed

Laser beam path (shown in red) if

New optic installed in "shutter enclosure" that motivated moving of laser safety "temporary" removal

Incident Description:
Laser operator noticed red laser beam on their shirt sleeve while moving a laser optic while lab in "Class 1" laser operation mode where laser beams are supposed to be fully enclosed.
No PPE laser eyewear was being worn, since not required for Class 1.

386

FIGURE 6.1   Optical set-up.

OPTION 2

Which is what the individual did, to his credit. He reported the occurrence to his supervisor, stating there was a safety lapse in their system that needed to be corrected. They should not have been able to, with no thought, remove the shutter that was a critical part of their safety system. These are the steps of a group that has a strong positive laser safety culture.

Are you asking for possible solutions? One, you could have a colored footprint of the shutter under the shutter, so when it is removed, there will be a clear sign or reminder that it is not in place. The second most common solution would be to secure the shutter with some special type of screw that requires a special tool to remove, with only a limited amount of people having access. I am sure you can think of other engineering controls.

The key to this near miss is that a safety flaw was identified and addressed in a formal and positive way.

SERVICE NEAR MISS

The case is rather simple. Per OSHA regulation, there is an interlock on a protective housing to be "bypassed," and that bypass will not allow the closing of the housing. The goal is to make sure the proper safety controls are put back in place at the end of service. Unfortunately, all repair persons and users have learned that the vast majority of protective housing interlocks can be bypassed with the use of paper clips or tape. Meaning that at the end of a long day, covers can be put back in place with the interlocks in a bypass mode. Meaning sometime later, someone can remove a cover counting on it to be depowered or beam blocked and receive an unfriendly shock, literally a reflected beam (see Figure 6.2).

WHY TRACK NEAR MISSES?

In simple terms, near misses are a warning that should be taken to heart. They usually represent an unsafe condition or practice that is uncorrected and can lead to a real

FIGURE 6.2 (a) Protective housing interlock, by passed, (b) Closer look at interlock taped down.

injury. Few of these are reported, investigated, or have corrective actions taken. Why? Because it was a near miss, no one was hit, someone got away lucky. Why cause trouble? Why admit a mistake? Why report? How are you going to feel when someone else is injured repeating just what you did, because they did not understand a risk or flaw existed.

Discussed below are additional near miss reports, covering procedural, technique, engineering, and human element issues—all of which reinforce the two examples used to start this chapter.

CASE 1: HUMAN ERROR AND RESPONSE

Occurrence Report Number: NA–SS–SNL–5000–2006–0004

The incident occurred at Air Force Facility on November 1, 2006, at 5:45 MST, involving Air Force and subcontractor personnel conducting a laser test.

At a staff meeting on November 2, 2006, the Department Manager was briefed on the previous night's test and told of an equipment anomaly during the test. Further information was requested from the Air Force operators. On Monday morning, November 6, 2006, the Project Manager, Principle Investigator (PI), and the ESH Coordinator reviewed the information and concluded that a laser beam release had occurred when subcontractors were in a nominal hazard zone (NHZ). It was concluded that this constituted a near miss event and reporting was immediately initiated.

Details

On November 1, 2006, at approximately 5:45 MST, two subcontractors were working in the experimental facility when the beam from the laser (Class 3B) was inadvertently released into the facility. The beam was released by an Air Force Safety Officer during a safety check, but as a result of not following the correct procedures. The Air Force Safety Officer was performing a shutter test, without a second beam block in place to prevent release of the beam. The beam block had been removed during the first safety check, and had not been replaced for the second test. The Air Force Safety Officer was new (first time working the experiment).

At the time of release, the subcontractors noticed a Charge-Coupled-Device (CCD) Camera behaving erratically. The subcontractors asked the Air Force Safety Officer about a potential release of the laser, to which the response was initially "no," but in fact, the laser had been accidentally released.

The subcontractors were later informed that there had been a temporary release of the beam. The subcontractors were not in the beam path; however, they were in the facility, which is included in the NHZ when the laser is present. The beam was shuttered within seconds of the release. They were not wearing the appropriate eyewear because they were not aware of the existence of the presence of the laser, and therefore were unaware of the resultant NHZ. The subcontractors were sent for a medical review of any potential exposure damage. They had luckily received no exposure damage.

Management Summary

During a laser test, a U.S. Air Force officer failed to follow procedures and inadvertently released a Class 3B laser beam within an experimental facility at the Air Force Base while two subcontract employees were working in the building. The two employees were not in the beam path, but were within the laser's Nominal Hazard Zone. The employees were not wearing laser eye protection, and were sent to the site medical facility for evaluation of any potential eye exposure damage. Although results were negative, it was determined that this event was a near miss and reporting was initiated.

The incident was discussed with the Air Force personnel. A test was run Thursday night, after assessing the situation, and taking the initial corrective action of replacing the Safety Test Director with a more experienced individual. They did not report the incident to OOPS at that point in time, mistakenly thinking that because it was an Air Force facility, and the Air Force was investigating, that they did not need to do so. After further consideration, they contacted others to seek guidance on whether they needed to report further or not.

Corrective actions:

1. Air Force Safety Test Director replaced with a more experienced Air Force personnel.
2. Individuals with potential exposure risk sent to medical to assess any health risks (eye damage).
3. Incident reports, and suggested improvements solicited from involved parties.
4. An initial suggestion to be implemented if possible is the addition of engineering controls in the form of an interlock switch controlled by experimental personnel. This will eliminate inadvertent releases prior to experimental staff being fully prepared, and mitigate potential human communication errors.
5. As a short-term solution, individuals will only open the laser port into the facility when necessary, and will wear goggles at all times while that port is open.

CASE 2: SEARCH FAILURE AND TOO CLEVER ENGINEERING

Occurrence report number SC–TJSO-JSA-TJNAF-2006-0005

At about 10:30 a.m., on Friday, December 1, 2006, at a Free Electron Laser Facility, free-electron laser (FEL) control room staff noted the presence of a worker in FEL Lab #1. Lab #1 had received less than 1 minute of FEL laser light at the time and the lab was immediately safed (laser light delivery stopped). FEL Lab #1 underwent a sweep to remove all workers to initiate "exclusionary state" (no one allowed in Lab #1) laser operations. The experiment configuration was such that the laser light was delivered in an isolated area and rope barriers were in place to prevent access to the experiment. The worker did not receive any injury.

Immediate Response

1. Within a minute of laser beam delivery to Lab #1, The FEL control room Duty Officer saw that the Lab #1 lights had been turned on and immediately safed Lab #1. This action released the magnetic Lab #1 door lock. The worker (a Hampton University technician supporting a laser experiment) then left Lab #1. There were no injuries to the worker and there were no ionizing radiation or radioactive/hazardous materials involved with this "near miss" event.

2. The Duty Officer then terminated all FEL operations and contacted the FEL Facility Manager. The FEL Facility Manager concurred that all FEL operations should cease. An electronic record was posted stating that no FEL operations were permitted.

3. The FEL Facility Manager immediately contacted the Associate Director for the FEL Division, Lab senior management, ESH&Q Division, and the DOE Site Office. An event investigation team was named and the team began event follow-up at 11:00 a.m.

4. The worker was examined at a clinic by the Lab physician during the afternoon. The examination verified that the worker had not received any injury as a result of this event.

Digging deeper here is a more complete incident summary:

The event investigation team's causal analysis determined that the event's direct cause was an FEL staff member's failure to do a thorough sweep of Lab #1 as required by the Laser Safety Operating Procedure. A properly conducted sweep requires an entire lab area search to verify no one is present. Contributing causes include the following:

1. New equipment had been recently added to Lab #1 that reduced the sweeper's field of vision.

2. The technician was using a computer that was not used in normal operations.

3. The sweep is time-limited to a 1-minute duration (by the safety system software), and this contributes to some urgency in getting the sweep done quickly.

4. Since Lab #1 was in an open state, and this worker was not involved with the lasing aspects of the experiment, no laser specific training was required. This lab-specific training instructs workers that the crash button on the Laser Personnel Safety System (LPSS) must be pushed as the exit button is inoperable in the exclusionary mode (this is the status immediately after the sweep and before/during laser delivery).

5. The exit buttons are not well engineered from a human engineering standpoint. There is no indication that they do not function when the Lab is an exclusionary mode.

Two phases of corrective actions are interim and permanent actions.

Interim Actions

1. Immediate FEL laser operations stand down.
2. Briefing of all Lab operational staff at the lab's two daily operations planning meetings. The outside FEL physics user group also receive a briefing on the event.
3. The FEL lab sweep procedure will be modified until permanent LPSS changes are in place. These interim changes include the use of two people for all pre-sweeps with no time limit. Once the pre-sweep is done, one person performs the sweep procedure called for in the procedure. During the sweep, the sweeper will push the mounted simulated sweep buttons that direct the sweep to all relevant parts of the Lab.
4. Personnel will be trained in the new sweep procedure.
5. Review this event with the users involved. The Jefferson Lab Director communicated the importance of safety, and the Lab's safety expectations to the principle investigators leading this experiment.
6. A sign will be posted near the exit buttons in all FEL labs indicating which button to use if the exit button does not work.
7. The Lab Laser Safety Officer (LSO) will brief all non-FEL Laser System Safety Supervisors on this event by December 11. This briefing will also determine if there are any extent of condition applications in other non-FEL laser activities.

Note: All interim actions were completed by December 4 except for action #7.

Permanent Actions

1. Two to four sweep buttons will be placed in each FEL lab to verify that the sweep goes to all portions of the lab. There will be a sweep button in each walk-in hutch as well.
2. There will be a verbal announcement that a sweep is occurring as the sweep begins.
3. There will be a verbal announcement that lasing is to begin after the sweep has taken place and 30 seconds before the beam is provided to the lab.
4. Ensure that FEL users are familiar with this event and applicable FEL protocols prior to initiating experiments.

More Background Details and the Free-Electron Laser (FEL)

The free-electron laser (FEL) is a high-power laser operating at wavelengths between 0.9 and 10 microns. It is tunable over each set of mirrors being used. The light is produced in a radiation shielded vault and transported through an evacuated transport system to one of seven user labs in the FEL facility. Each lab can be swept (cleared of all personnel), locked up, and brought up in a state of "Laser Permit." If an interlock is tripped or a crash button is pushed, the lab is *crashed* and transitions to an *open* state. If an open state is required, the FEL laser operator can *safe* the lab using the laser control system. When the lab is in an open state, access is permitted to anyone and the only training required is the training required to be on the accelerator site. When one wants to put the lab into a Laser Permit state, an approved laser user or FEL laser operator sweeps the lab according to a sweep procedure documented in the Laser Standard Operating Procedure. There are three possible states in a lab when a lab is in Laser Permit. In *Exclusionary* mode, no-one is permitted in the user lab. This is the default state for a lab. In *Hutch* mode, the laser is confined to interlocked hutches inside the lab. In *alignment* mode, the power to the lab is restricted to low levels so that a user with laser safety eyewear can align the laser beam to their apparatus. Trained personnel (identified by an RFID badge) can enter the lab when it is in alignment mode (if they are wearing appropriate laser safety eyewear) or in hutch mode. No access is permitted in exclusionary mode. Egress from the lab is accomplished by pushing an exit button on the door of the Laser Personnel Safety System (LPSS). These buttons only function in alignment mode or hutch mode. To gain egress in exclusionary mode, it is necessary to use the crash button on the LPSS chassis. This is taught to all approved laser users.

A sweep consists of presenting one's RFID card, entering the lab, verifying that all personnel have exited the lab, pushing an "initiate sequence" button, exiting the lab and presenting one's RFID card a second time. In this case, the lights were turned off after pushing the initiate sequence button. Thirty seconds after pushing the initiate sequence button, the lab goes into a state of Laser Permit. During this time, a warning beacon is illuminated and a warning alarm is sounded. When the lab is in laser permit, the warning alarm is turned off and the warning beacons stay on.

First, the lab was locked up and safed several times during initial checkout of the experiment's diagnostics.

Subsequently, it was determined that a camera had to be replaced so the lab was safed and three technicians and one of the FEL Operators entered the lab to change it out. During the entry, a user associated with the LIPSS experiment went into the lab to use a computer along the North Wall to check his email. Once the camera was replaced, the three technicians left the room and the FEL Operator swept the lab in order to bring it into laser permit. This was at about 10:28 a.m. The start of the sweep was announced by the Operator, but the user did not hear him. The FEL operator did not do a thorough sweep and did not see the technician. Following the sweep, he pushed the initiate sequence button, turned off the lights, and exited the lab. The user realized that the lab was going into laser permit and tried to leave the lab using the exit button on the door to leave.

When the lab is in exclusionary mode, the exit buttons are nonfunctional. Instead, it is necessary to hit the crash button to leave the lab in exclusionary mode, so pushing the exit button did not release the mag locks on the door, and the user was unable to exit. In order to notify the control room that he was in the room, he turned the overhead lights on and off and then left them on, with the intention of bleaching the camera response to alert the control room staff that there was a problem. He also ensured that he remained at the door facing with his back to any potential stray laser beams. The user had the required Laser Safety Orientation (SAF 114O) and Laser Medical Approval (SAF 114E) in order to be a user. He did not have, nor was he required to have, laser specific training to be in the lab when in the open state.

FIGURE 6.3   View of Lab #1 from entrance door. Computer workstation is located along the wall on the right side of the lab, behind the magnet set-up.

The FEL operators sent 25 W of FEL light at 990 nm into the lab. The light was directed through the optical transport onto a power meter for a total of 1 minute (see Figure 6.3). When the FEL operator saw that the lights had been turned on in the lab, he immediately safed the lab. This was at 10:30 a.m. When the lab was crashed, the technician was able to immediately exit the lab.

The Duty Officer then terminated FEL operations and contacted the facility manager. He concurred that all lasing operations should cease, and posted an electronic record called an FLOG indicating that FEL operations were not permitted. He then contacted the AD, the lab laser safety officer, and the laser system supervisor.

Lessons Learned, Including EH&S Manual Changes (If Any)

A persistent problem with administrative controls in safety systems is that their effectiveness depends on the diligence of the person enforcing the control. The importance of this diligence must be continually reinforced over time so that incidents like this do not occur. Engineering controls can force a sweeper to take their time and cover the required territory but are no substitute for being careful in the sweep. It is also important to continually stress that personnel take whatever time is required to do a careful sweep and not rush the sweep.

Another lesson learned is to consider human engineering when designing the user interface for a safety system. Inexperienced users are reticent to push crash buttons since they think they will break something. They would rather push an exit button. If an exit button is present but nonfunctional, they will typically be confused and not know what to do. The best solution is for the exit button to be the crash button.

Finally, it is important to consider how things work when things do go wrong. Even if an event is unlikely, it is important that things work as they should during the event.

CASE 3: LEGACY AND FUNDING EVENT

During the installation of a new laser interlock system at the ATF, some potential safety risks were discovered in the previously existing system. The interlocks are very complex as there are two Class 4 lasers, whose outputs are sent to different places in order to excite the photocathode that provides electrons for the Linac. They are used in different rooms as part of the various experiments and there are many different configurations on the beam lines. The complexity is reflected in the test procedures where there are 246 steps, many with multiple-step check-offs.

The project had been transferred between groups with varying funding resources. It was determined that both the radiation and laser interlocks were beyond the expertise of the new group and required outside expertise. A memoriam of understanding was developed:

> C-A Department has the responsibility to maintain or modify the ATF access control system hardware and maintenance procedures. This includes, but it is not limited to, logic diagrams, wiring diagrams, test procedures, interlocks, interlocked gates, beam crash buttons or cords, reset stations, critical devices, and Radiation Safety Committee review of same.

In order to facilitate this, a complete set of drawings for both systems were requested and received. A management review led to the following memo:

a. **Safety Systems**
   i. **Laser Interlocks:** The ATF has a complex set of interlocks that has been weakly supported by the institution interlock group leaving the ATF with a system that is not properly documented for the system that is in place. ***Changes over the years are not documented and the procedures for testing do not properly cover the current configuration. Furthermore, those that designed and implemented the system are refusing to support the ATF even as purchased services. This is a critical situation.*** We are actively trying to get the C-A Department to review and take responsibility for the maintenance and testing of the ATF interlocks but the cost of redoing the system may be prohibitive.

Testing and re-certify the laser interlock system by adding to the test procedures the items known not to have been included, believing the rest of the test procedures to be complete. At the time, ATF did believe that they could not verify the completeness of the test procedures because they were concerned they never had complete drawings. At the time of the turnover from the one group to another department, the previous engineer stated that all the changes to the drawings had been completed but changes to the test procedures had not. The changes to the test procedures were completed in July 2004. The system continued to operate believing that all was in order. However, this could not be independently verified. To this end, it was decided that the system would have to be replaced, completely redocumented, verified, and certified. Monies were allocated for this and with the help of C-AD, was completed in June 2007.

Interlock Tests

ATF procedures, in compliance with the SBMS subject area, require laser interlocks to be tested every 6 months. Original checklists are kept in the ATF Control Room with copies sent to the Department's Safety and Training Office where the Manager of the ESH&T program reviews them.

The test procedures were originally developed by the institution and since (working with the C-AD) have been modified to remove testing for devices no longer physically present, added testing for a device added into the interlock chain, and redefined in some areas (two separate areas merged into a single area). At that time, the Department Chair requested assistance from the institution Chair who permitted their interlock engineer (person involved in the original design, implementation, and testing) to help facilitate the changes. When the modified procedures were acceptable to both the groups and C-AD, they were adopted and published.

The Concerns That Arose

In replacing the old laser interlock system with the new one, the plan was to replace the Programmable Logic Controller (PLC) with the same type currently used by the C-AD interlock group for their interlocks in order to use the same programming language and take advantage of their expertise with this hardware and software. In principle, the new controller boxes had the same inputs and outputs as the old ones, so replacing the new boxes compatible with the new PLC and language should have been easily done and no new wiring was expected. The old bundle of wires simply had to be plugged into the new boxes. After this was done, the system began to be checked and discrepancies were found.

- It was found that the drawings as given from the institution were not complete or accurate. ATF personnel found approximately 30 errors on the existing drawings.
- Additionally, it is doubtful that a full functionality test had been done on the status indicator boxes as there was an LED that was wired backward and could never have worked.
- In certain cases, an indicator light, that implied *two* shutters were closed, was lit when only one shutter was actually closed. Other laser status panel *on* LEDs could not be correctly controlled because they were hard-wired to a *shutter open* switch. In order for the LEDs to be correctly controlled, they need to be independently driven by the interlock system.
- The YAG shutter status LEDs on the CO_2 room enclosure did not reflect the position of the YAG shutter as the LEDs were not correctly wired. The shutter *open* LED would light whenever the CO_2 to high-bay pass-through button (means to enter the interlocked area by authorized personnel) was pressed. The shutter *closed* LED would light when the 10 *micron imminent* LED was on. While these wiring problems did not add additional risk to personnel, they are indicative of the lack of a full functional test.

- Finally, it was apparent that the normal "best practices" for color coding different sets of wires was not done as most of the wiring was the same green color.

Safety Implications

The improper lighting of status indicators, which were viewed by users in order to determine which laser eyewear to use, put them at potential risk for eye damage. This authorized user list was limited in number to those specifically included in the approved Experimental Safety Reviews. These people had the proper training, including the initial eye examinations given to authorized laser users. They are not permitted to do laser alignments or maintenance. Their use of laser light was limited to their experimental equipment.

The incorrect indicators of the laser safety shutter's state put users and staff at additional risk. Prior to the transfer of the ATF to the department, the institution was responsible for the design, documentation, installation, programming, and testing of the laser interlock and radiation systems. In the past the institution ran functionality tests of the whole system, apparently it seemed that rewiring must have been done at some point afterward, as the backward wiring would have been found at that time as well as the false indication that two shutters were closed when in fact the indicator light was lit when only one was closed.

An interviewed user stated "I did not know that the full interlock test procedures did not cover each and every item in the system and that it took 5–6 days to do. I was informed this morning that he didn't know this himself until last week. I presumed that when there was a modification the interlock test procedures tested the whole system."

Analysis to This Point

1. Unlike the interlocks for radiation, the areas covered by laser interlocks alone are not subject to the Accelerator Safety Order.
2. Unlike the area encompassed by the interlocks for radiation (which is an exclusion zone when radiation is present in the Experimental Hall) the laser areas are configured for occupation by properly trained persons (with the proper PPE) when the interlocks are on and laser light is present in the rooms. This allows properly trained people to make adjustments, align the lasers, maintenance, and approved experimentation.
3. The status indicator lights are not required but are part of the interlock safety system as they get their information from the interlock circuitry and users will usually use this information to choose the correct eyewear.
4. Laser safety eyewear is intended as an additional control to reduce risk for eye damage and is required by the ANZI Standard. The laser SOP is used to analyse the hazard and specify the optical density. All of that is in order.
5. There is no reason to believe that any user received any eye damage, although the potential was there.
6. It is prudent to see if the SDL has a similar situation. Have they had a full functionality test? Do they rely on indicator lights for eyewear? If so, is that system working properly?

7. Looking at the Categorizer's Procedure, it might come under the following but is at the discretion of line management and not required to be reported:

Group 10 - Management Concerns/Issues

(2) 1–4† An event, condition, or series of events that does not meet any of the other reporting criteria, but is determined by the Facility Manager or line management to be of safety significance or of concern to other facilities or activities in the DOE complex. One of the four significance categories should be assigned to the occurrence, based on an evaluation of the potential risks and the corrective actions taken.

(3) 1–4† A near miss, where no barrier or only one barrier prevented an event from having a reportable consequence. One of the four significance categories should be assigned to the near miss, based on an evaluation of the potential risks and the corrective actions taken.

Present Status

There is a new interlock system in place. This system is fully documented with a new set of drawings, the hardware has been installed by ATF personnel, the controlling software was written by the C-AD interlock group, tested by C-AD interlock technicians, and was reviewed and certified by C-AD personnel. The status indicator lights are currently not part of the system but will be added to it, documented, and test procedures will include their testing. Until then, no users will be permitted to be in the laser interlocked areas until the status lights are operational and certified to be working properly.

For the future, any modifications will require a full functional test as well as properly documented updates to drawings.

CASE 4: DELIBERATE BYPASSING OF A SAFETY SYSTEM

While performing a series of integrated dry runs, in October 1998, the personnel discovered that three interlocks on the experiment room door were taped in a bypassed condition. The preliminary inquiry concluded that the bypassing might have taken place the preceding evening, when site personnel were ensuring that two Class 4 lasers were correctly aligned for the integrated dry runs. The test group director prohibited the operations of x-ray and laser equipment until the incident was fully evaluated. If such equipment is operating while an interlock is overridden, entry into the area will not shut the equipment down, and personal injury could occur.

A follow-up investigation determined that facility workers attempting to identify which interlock was connected to the laser systems had systematically taped the door interlock devices in a bypassed condition, in violation of the approved operating procedure. It was also determined that the facility staff forgot to remove the tape before the dry runs. The review concluded that since the experiment room door had cipher locks and access was administratively controlled, there was no chance of personnel exposure of injury.

Case 5: Wanting to Be Helpful But Just Following through and Not Thinking

In September 2006, a plumber asks a researcher (visiting scientists) to let him into an interlocked laser lab. The researcher puts in the access code, allowing the plumber to enter, and then returns to his office. Oh yes, the lasers were running in the laser lab. The PI for the lab is in the lab doing a laser alignment. He sees the plumber (without eyewear) and escorts him out.

Case 6: Near Miss at Los Alamos National Lab

In September 2000, a service man is working on an LSR flow cytometer (9/20). These units contain between 2 to 3 class 3B lasers. He removes the top enclosures and housing over the lasers. The service man leaves the unit in this way and plans to return in a few days to complete the repair. Three days later, a Group Leader observes two employees standing over the unit, looking inside of it with the laser on.

NEAR MISS REPORTING

A thorough near miss investigation has the potential to identify overlooked physical, environmental, or process hazards, the need for new or more engineering controls, or safety training, or unsafe work practices. The supervisor of the employee involved in the near miss is responsible for conducting the investigation and, when appropriate, ensuring that corrective actions are taken. The depth and complexity of the investigation will vary with the circumstances and seriousness of the incident. Investigators must maintain objectivity throughout the investigation. The purpose of the investigation is to uncover any factors that may have led to the accident, not to assign blame.

The first is all narrative:

Background Information: Site location, department, supervisor, date, time, type of work

Witnesses: Self-explanatory

Description of Accident: Most near misses result from an accumulation of events. An accurate, factual description of the accident and the events leading up to it can be very helpful. This chronological sequence can be studied to determine how each event may have contributed to the near miss. Include photos or drawings of the accident site, if these will be useful to the investigation.

Factors: Factors, if any, are the conditions in the workplace or actions that contributed to the near miss of this near miss. Examples might include unguarded machinery, broken tools, and slippery floors, not following established procedures, or insufficient training or maintenance.

Corrective Actions: List actions or steps that could be taken to control or eliminate the likelihood of a recurrence. Include not only those that can be accomplished right away (for example, providing personal protective equipment, installing a machine guard), but also actions such as changes in policy or providing additional training.

Near-Miss Report Form Sample

DEPARTMENT	PHONE #	Environmental Health & Safety		
1. EMPLOYEE *(last name, first name, mi)*	2. EMPLOYEE ID No.	3. SEX ☐M ☐F	4. AGE	5. DATE/TIME OF INCIDENT
6. TIME IN JOB ☐ Less than 1 mo. ☐ 6 mos. to 1 year ☐ 1 to 5 mos. ☐ More than 1 year	7. JOB TITLE AT TIME OF ACCIDENT	8. EMPLOYMENT CATEGORY ☐ Full-time ☐ Temporary ☐ Part-time ☐ Student		
9. SPECIFIC LOCATION OF NEAR MISS *(bldg., floor, room #, outside)*		10. WITNESS *(list name(s) & phone #)*		

11. DESCRIPTION OF NEAR MISS *(Describe sequence of events, including time, date, and location of incident. Attach photos, drawings, or separate page if necessary)*

12. FACTORS (Why It Happened) *Describe conditions or practices, if any, that may have led to the occurrence of this incident. Attach separate page if necessary*

13. CORRECTIVE ACTIONS (Prevention). *Developed jointly with EH&S*

14. REPORTED BY	16. EH&S REVIEW
_____ Signature Date _____ _____	_____ _____ _____ _____
15. DEPARTMENT HEAD/SAFETY MANAGER COMMENTS. _____ _____ _____ _____ _____ Department Head/Safety Manager Signature Date	_____ _____ _____ _____ _____ Signature Date

THE PERCEIVED INCIDENT

Tying in with the near miss is the perception of risk. This must be dealt with as if real. First, the LSO wants to make sure there are no hazards, second, we want to make staff feel reassured they are safe, and lastly, if perception is treated poorly or the reporter is made to feel foolish, this can cut down on reporting. Which might prevent a real hazard from coming to your attention.

You do not want unaddressed perceptions to be sent to regulatory agencies which could open up into a bigger problem.

CONCLUSION

The near miss incident can easily show problems with one's safety culture or that of others. Therefore, the reporting of near miss incidents should be supported and employees encouraged to do so. A hard goal to accomplish, but it has a positive pay off.

7 Where Do I Find Laser Accident Information?

Ken Barat

CONTENTS

The importance of sharing information on laser accidents with your user population cannot be understated. But a common question from laser safety officers is outside of something that has happened at my facility: How do I learn about such incidents?

The question is extremely valid. It is a challenge to find creditable data on laser accidents and then to turn that information into a lesson learned that will have meaning to your staff. However, a number of sources do exist that are worthwhile for one to check on, periodically, for such information.

DOE EFCOG LASER SAFETY WORKING GROUP

The most readily accessible source of laser lessons learned is from the DOE EFCOG Laser Safety Working Group. Laser incidents are of great concern to the U.S. Department of Energy. Each incident gets an investigation with a written report. Information from those reports in a summary format can be found of the website of the DOE EFCOG Laser Safety Working Group.*

---

* EFCOG, *Laser Safety Task Group/LLNL Laser Lessons Learned Newsletters*, http://efcog.org/safety/worker-safety-health-subgroup/laser-safety-task-group.

The Laser Safety Working Group is one of several Subgroups of the Environmental Safety & Health Working Group of the DOE Energy Facilities Contractors Group (EFCOG) Mission. The Laser Safety Subgroup promotes excellence in all aspects of laser safety through the collaborative sharing of policies, best practices, written procedures, tools, lessons learned, and hazard control technology. This Subgroup establishes an effective network for laser safety experts from DOE facilities: (1) Vision Laser work will be done safely at DOE facilities with minimal risk of hazardous exposure, (2) laser workers will have excellent laser hazard awareness and the necessary skills, equipment, and tools to mitigate this hazard, (3) affected personnel will also have an appropriate level of laser hazard awareness, and (4) the Laser Safety Subgroup will contribute to improvements in practices, policy, education, and training for laser safety.

In addition to the lessons learned on this website, one can access the LSO Workshop presentations.* As you look at the LSO Workshop, one will find presentations on some individual laser incidents. For those not familiar with the DOE LSO Workshop, it is for individuals with laser safety responsibility and interest in a research, industrial, or academic setting who want to update and expand their laser safety knowledge. It features presentations on current laser applications and associated laser safety issues and solutions. Anyone interested in laser safety will not want to miss this unique workshop! The workshop also serves as the official annual meeting of the U.S. Department of Energy Laser Safety Task Group (DOE–EFCOG).

MAUDE-CDRH-FDA

The MAUDE database is freely accessible to anyone. One can go to the FDA or CDRH website and search for databases to find the current link to this resource. The following is a detailed explanation of what MAUDE is and why it exists.

Each year, the FDA receives several hundred thousand medical device reports (MDRs) of suspected device-associated deaths, serious injuries, and malfunctions. The FDA uses MDRs to monitor device performance, detect potential device-related safety issues, and contribute to benefit-risk assessments of these products. The MAUDE database houses MDRs submitted to the FDA by mandatory reporters (manufacturers, importers, and device user facilities) and voluntary reporters such as healthcare professionals, patients, and consumers.

Although MDRs are a valuable source of information, this passive surveillance system has limitations, including the potential submission of incomplete, inaccurate, untimely, unverified, or biased data. In addition, the incidence or prevalence of an event cannot be determined from this reporting system alone due to potential under-reporting of events and lack of information about frequency of device use. Because of this, MDRs comprise only one of the FDA's several important postmarket surveillance data sources.

---

* EFCOG, *DOE Laser Safety Officer Workshops*, http://efcog.org/safety/worker-safety-health-subgroup/laser-safety-task-group/doe-laser-safety-officer-workshops.

- Please note that the MAUDE web search feature is limited to adverse event reports within the past 10 years.
- MDR data alone cannot be used to establish rates of events, evaluate a change in event rates over time, or compare event rates between devices. The number of reports cannot be interpreted or used in isolation to reach conclusions about the existence, severity, or frequency of problems associated with devices.
- Confirming whether a device actually caused a specific event can be difficult based solely on information provided in a given report. Establishing a cause-and-effect relationship is especially difficult if circumstances surrounding the event have not been verified or if the device in question has not been directly evaluated.
- MAUDE data does not represent all known safety information for a reported medical device and should be interpreted in the context of other available information when making device-related or treatment decisions.
- Variations in trade, product, and company names affect search results. Searches only retrieve records that contain the search term(s) provided by the requester.
- Submission of a medical device report and the FDA's release of that information is not necessarily an admission that a product, user facility, importer, distributor, manufacturer, or medical personnel caused or contributed to the event.
- Certain types of report information are protected from public disclosure under the Freedom of Information Act (FOIA). If a report contains trade secret or confidential business information, that text is replaced by "(b)4)". If a report contains personnel or medical files information, that text is replaced by "(b)(6)". The designations "(b)(4)" and "(b)(6)" refer to the exemptions in the FOIA. For example, "(b)(4)" may be found in place of the product's composition and "(b)(6)" may be found in place of a patient's age.
- MAUDE is updated monthly and the search page reflects the date of the most recent update. The FDA seeks to include all reports received prior to the update but the inclusion of some reports may be delayed.

ROCKWELL LASER INDUSTRIES (RLI)

Rockwell Laser Industries is a long-established laser safety consulting firm, started by one of the fathers of laser safety in the United States, Jim Rockwell. The laser accident database is maintained by RLI. As with any database, it is only as good as its data. Any case entered into the RLI database must be verified; either by the facility where the accident occurred or by a credited source, such as a regulatory agency.

The laser accident database has been gathering data since 1964 and is free to access by going to the RLI webpage. Below are some summary charts from previous data within the database (Figures 7.1 through 7.3).

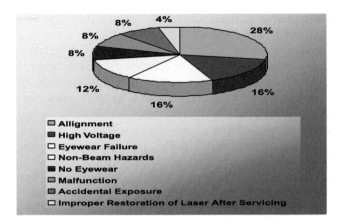

FIGURE 7.1 Accident settings.

Number of Incidents by Environment
First 25 Years vs. Last 25 Years

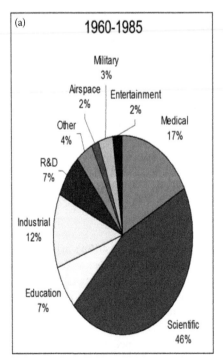

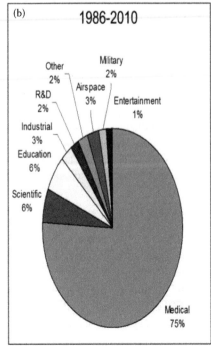

FIGURE 7.2 (a) Incidents by environment 1960–1985, (b) Incidents by environment 1986–2010.

Number of Incidents by Laser Type
First 25 Years vs. Last 25 Years

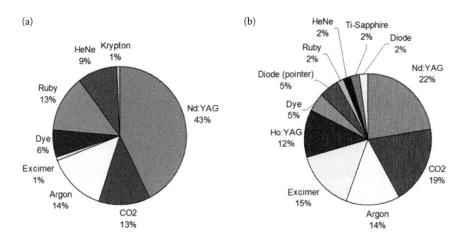

FIGURE 7.3 (a) Incidents by laser type 1960–1985, (b) Incidents by laser type 1986–2010.

OCCUPATIONAL SAFETY AND HEALTH ADMINISTRATION (OSHA)

With the Occupational Safety and Health Act of 1970, Congress created the Occupational Safety and Health Administration (OSHA) to assure safe and healthful working conditions for working men and women by setting and enforcing standards and by providing training, outreach, education, and assistance. The OSH Act covers most private sector employers and their workers, in addition to some public-sector employers and workers in the 50 states and certain territories and jurisdictions under federal authority. Those jurisdictions include the District of Columbia, Puerto Rico, the Virgin Islands, American Samoa, Guam, Northern Mariana Islands, Wake Island, Johnston Island, and the Outer Continental Shelf Lands as defined in the Outer Continental Shelf Lands Act.

REPORTS AVAILABLE FROM OSHA

Workplace Injury, Illness, and Fatality Statistics

Injury/Illness Incidence Rates

- Industry Injury and Illness Data
- State Occupational Injuries and Illnesses

Injury/Illness Characteristics

- Case and Demographic Characteristics for Work-Related Injuries and Illnesses Involving Days Away from Work

Fatalities

- OSHA Weekly Fatalities and Catastrophes (FAT/CAT) Reports
- BLS Census of Fatal Occupational Injuries, 1992–Present

BLS (Bureau of Labor Statistics)

- BLS Safety and Health Statistics Home Page
- Keyword Search of Available BLS Injury/Illness and Fatality Data, and Publications

AND NOW FOR SOMETHING DIFFERENT: THE DARWIN AWARD

The Darwin Award started in 1993 and is named after Charles Darwin and his theory of survival of the fittest. The description of the Darwin Awards states:

> The Darwin Awards commemorate individuals who protect our gene pool by making the ultimate sacrifice of their own lives: by eliminating themselves in an extraordinarily idiotic manner, thereby improving our species' chance of long-term survival. In other words, they are cautionary tales about people who kill themselves in really stupid ways, and in doing so, significantly improve the gene pool by eliminating themselves from the human race. These individuals carry out disastrous plans that any average pre-teen knows are the result of a really bad idea. The single-minded purpose and self-sacrifice of the winners, and the spectacular means by which they snuff themselves, make them candidates for the honor of winning a Darwin Award.*

Why mention such a source? For, as we all know, sometimes humor is the best way to get a message across or to get people's attention. So that is the explanation of why the Darwin Awards is included in this chapter. At times, a clever safety person can take the incident in a Darwin Award and relate it to actions or activities at their own facility or user population.

FEDERAL AVIATION ADMINISTRATION (FAA): COLLECTING DATA ON LASER INCIDENTS

The FAA focus is on collecting data about laser incidents, primarily pilot illumination. Which is a major problem worldwide. The best items from a lessons-learned perspective are the free downloadable videos on the FAA website, www.faa.gov, on the problem of pilot illumination. These videos are great for a safety meeting. While the topic might not directly seem to relate to one's work situation, they can be the jumping off point for conversation on improper use of lasers, bioeffects, user responsibility, and so on. The rest of the FAA website is more of a request to obtain information on incidents. In the FAA's own words "In cooperation with federal, state and local law enforcement agencies, FAA needs everyone's help in reporting laser incidents. If you are the victim of a laser incident or you witness a laser incident, please report it to FAA." Here's how:

---

* The Darwin Awards, *Rules: What Are They?* http://www.darwinawards.com/rules.

PILOTS AND CREWMEMBERS

Per FAA Advisory Circular (AC) 70-2A, *Reporting of Laser Illumination of Aircraft*, all pilots and crewmembers are requested to immediately report incidents of unauthorized laser illumination by radio to the appropriate ATC controlling facility.

Upon arrival at destination, all pilots and crewmembers affected by an unauthorized laser illumination are requested to complete the FAA Laser Beam Exposure Questionnaire in order to provide critical information in support of law enforcement efforts to identify and apprehend the responsible parties.

- Complete the short version FAA Laser Beam Exposure Questionnaire on your mobile device. We will then email you the full questionnaire for you to complete and return with additional information.

- or -

- You can download and complete the FAA Laser Beam Exposure Questionnaire (PDF) from your personal computer. Completed questionnaires can be saved and attached to an email sent to laserreports@faa.gov, or can be printed and faxed to the Washington Operations Control Center Complex (WOCC)— (202) 267-5289 Attn: Domestic Events Network (DEN).

AIR TRAFFIC CONTROL

Submit laser incident information to the Domestic Events Network (DEN). For guidance, see *AC 70-2A, Reporting of Laser Illumination of Aircraft*.

THE PUBLIC

If you're a member of the public who witnessed an individual aiming a laser at an aircraft, send an email to laserreports@faa.gov and include the following information:

- Your name and contact information
- Date and time you witnessed the laser incident
- Location and description of the incident

After the FAA has received your email, FAA staff or the appropriate law enforcement agency may decide to contact you if additional information or clarification is needed.

DATABASES THAT EXIST, BUT HAVE LIMITED OR CHALLENGING ACCESS

LASER ACCIDENT AND INCIDENT REGISTRY (LAIR) LAIR

LAIR is sponsored by the Laser Bioeffects Branch of the U.S. Army Medical Research Detachment (USAMRD), Walter Reed Army Institute of Research, a major part of the Tri-Service (Army, Navy, and Air Force) Directed Energy Bioeffects Research Center at Brooks City Base, Texas. USAMRD began collecting laser incident information in the early 1970s, and they now have more than 320 incidents on file, including military accidents investigated by their in-house teams, and nonmilitary incidents collected from outside sources. LAIR is a research-oriented database used to support U.S. military operations through the practical application of information extracted from laser accidents and incidents. LAIR is accessible on the internet and contains comprehensive data fields for documenting the laser incident and subsequent management of any injuries. USAMRD is the hub for all laser injury evaluations conducted by the U.S. Department of Defense, in addition to conducting a vigorous laser bioeffects research program. USAMRD has established and maintains unique expertise in investigating, evaluating, and treating laser eye injuries. Their multidisciplinary investigation team is comprised of laser systems and bioeffects experts, psychologists, and an ophthalmologist, all with extensive laser incident investigation experience. All three military services have established policy recognizing the importance of their respective Brooks City Base organizations in the investigation of laser incidents. Army policy (Army Regulation 9-11), Navy policy (Bureau of Medicine and Surgery Instruction 6470.23), and Air Force policy (Air Force Occupational Safety and Health Standard 48–139) require or strongly encourage notification of their components at Brooks City Base when laser exposure incidents occur.

Of the 29 military laser injury reports in the unified database from 1965 to 2002, there were 6 Air Force, 15 Army, and 8 Navy/Marine injuries.

U.S. DEPARTMENT OF DEFENSE (DoD)

U.S. DoD has recognized the need for a general guidance for laser incident reporting within the DoD, the Laser System Safety Working Group, which includes laser experts from all services, drafted a new *DoD Instruction for Protection of DoD Personnel from Exposure to Laser Radiation and Military Exempt Lasers.*

This instruction directs all services to report suspected laser exposures to the tri-service laser hotline at Brooks AFB. *Instruction 6055.15: DoD Laser Protection Program*: Implements policy, assigns responsibilities, and describes procedures in support of DoD laser protection. Establishes the DoD Laser Systems Safety Working Group (LSSWG), the Tri-Service Laser Injury Hotline, and the Laser Accident and Incident Registry. Requires that DoD Components that own and operate lasers shall establish and maintain a laser protection program that conforms to the requirements and procedures in this Instruction.*

Therefore, if you have access to this database, it is another source for lessons learned.

AVIATION SAFETY REPORTING SYSTEM (ASRS)

ASRS is a voluntary, confidential, and anonymous incident reporting system funded by the FAA but administered by the National Aeronautics and Space Administration

---

* Homeland Security Digital Library, *6055.15: DoD Laser Protection Program*, https://www.hsdl. org/?abstract&did=473728.

(NASA). Its purpose is to identify aviation hazards and safety discrepancies in support of aviation safety research and policy formulation. Pilots, air traffic controllers, flight attendants, mechanics, and ground crew can report any incident where they feel aviation safety was compromised by submitting the ASRS forms to the National Aeronautics and Space Administration. The ASRS is accessible on the internet and contains reports collected since 1988. The database contains a large number of data fields that are nearly identical to two other databases, the National Transportation Safety Board (NTSB) Aviation Accident/Incident Database and the FAA Incident Data System. The fields permit the input of extensive information about the circumstances of the incident, including location, time, date, weather conditions, flight plans, description of nearest airport, aircraft description, and pilot experience. The database supports textual search of all data fields. On March 4, 2003, a search for laser incidents turned up 16 relevant incidents out of 266,076 reports. An injury was reported in 3 of the 16 laser exposure incidents: 1 case of multiple flash burns to the cornea, 1 burned retina, and 1 case of conjunctival burns. In all three cases, the circumstances of exposure and the reported injury are incongruous. Other interesting reports included two incidents where the laser purposely tracked the aircraft, two incidents where an attendant took a laser pointer away from a passenger who was purposefully shining the laser in people's faces, and one incident where an air traffic controller erred in routing an aircraft around an outdoor laser system, resulting in a violation of minimum aircraft separation rules.*

OTHER DATABASES

I am sure there are other databases on the internet, such as the National Fire Protection Association and other organizations for members to report laser fires, but I am unaware of any reported with ease of accessing their data. Other databases may be hosted by private individuals. The ones listed above are the ones I am most aware of. Most importantly is the fact that such databases exist and can be used to reinforce good practices and highlight the results of poor practices or equipment malfunctions.

---

* Lt Col Lawrence K. Harrington and Jeffrey C. Wigle, "Ocular Laser Exposure Incident Reporting," *Military Medicine* 169, no. 4 (2004): 277.

8 R&D Accident Case

Ken Barat

CONTENTS

This incident happened at a Department of Energy research lab. Some of the take aways from this incident are of greater importance than the accident itself. They are:

- Failure of on-the-job training/mentoring
- Failure to make people aware of policies
- People will always do things you could never imagine them doing; in other words, there is no limit to stupid actions

So, let's start with the accident and then go back to the three items above.
Basic Facts:

1. A student with little ultrafast laser experience is working in a lab as a user
2. Not a member of the research group
3. Receives 15 min of on-the-job training
4. Performs an 800 nm alignment activity without the use of available eyewear (this is how he has observed others working)
5. Key optic is rotating the polarizer, a reject beam
6. Has no direct supervisor
7. Due to item 6 and 2, no one reviews his work plan
8. When alignment is completed, he rotates the tube with the beam-splitter attached to the polarizer to line up one hash mark with zero on the polarizer, just because he thinks it looks better that way

Investigation and additional background material (Figure 8.1):

- Polarizer-1 with escape window
- Beam tube with escape window
- Rotating mount

77

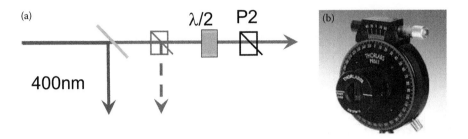

FIGURE 8.1   (a) Critical part of optical path, (b) Rotating polarizer.

Initially, the beam tube was secured to block the o-ray, which is not used in the experiment.

The laser operator wanted the P1 polarizer to be orthogonal to the incident laser beam and removed eyewear to co-align back-reflected beam from polarizer with incident beam. To do this, the operator loosened the polarizer and associated beam tube in the rotating mount.

The operator then wanted to get the polarizer rotation angle oriented to match the 0-degree marking on the rotation mount. Accidental exposure happened when the operator rotated the polarizer in a way that the escape windows of the polarizer and beam tube aligned, permitting the o-ray to strike their unprotected eye.

If warning signs were enough to yield compliance, this lab would not have had any issues. In addition, it also proved a point of mine, door interlocks do nothing to provide safety to or for authorized users. They are to protect those who should not be there (Figures 8.2 through 8.5).

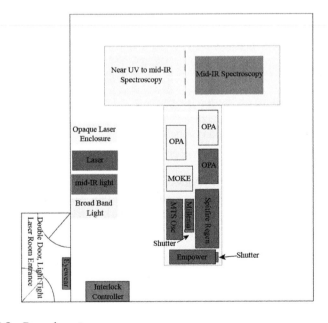

FIGURE 8.2   Room layout.

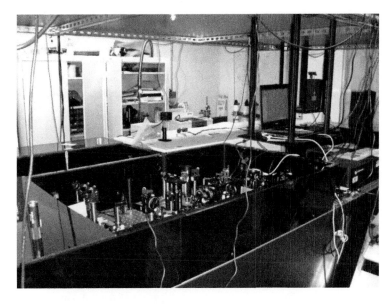

FIGURE 8.3   Open path portion of set-up.

FIGURE 8.4   Inside door.

FIGURE 8.5   Entry lab door.

POSSIBLE SOLUTIONS TO THIS INCIDENT

ENGINEERING CONTROLS

Use a beam tube with no escape window. Avoid mounting polarizers in rotating mounts, unless absolutely necessary. Instead, use half waveplates to rotate the polarization vector and keep both transmitted and reflected beams in a horizontal plane.

ADMINISTRATIVE PROCEDURES

For aligning the polarizer normal to the beam, the user can use three methods with full protection eyewear:

- IR sensor card with a hole in it; incident beam goes through hole and view back reflection on sensor card as polarizer pitch and yaw are adjusted.
- Cut an IR card so user can have a straight edge of sensitive IR card close to the beam.

- Use an iris and IR viewer to see back reflection on the iris.
- For performing optics adjustments when there's no need to observe the beam, block the beam upstream.
- Not necessary to have polarizer axes oriented at 0-degrees on rotation mount, so don't perform unnecessary optics adjustments where laser beams are present.

PPE

Use of full protection eyewear as required by the SOP

Accident investigation came up with seven root causes:

1. Inadequate training, in particular on-the-job training (OJT)
2. Inadequate supervision
3. Inadequate work planning and control
4. Inadequate adherence to requirements
5. Deceptive hazard of a dimly visible beam
6. Inadequate appreciation of out-of-plane beam hazard in using a polarizing beam-splitter
7. Inadequate intervention following (prior) PPE safety violations

Item 6 is so important that more needs to be said, since it has been at the center of many a laser injury. Polarizing beam-splitters are designed to split light into reflected S-polarized and transmitted P-polarized beams. They can be used to split unpolarized light at a 50/50 ratio, or for polarization separation applications such as optical isolation. High-performance polarizers are based on birefringent crystals.

Another fact is that 100 mW at 700 nm (nearly visible) will appear to have similar brightness as 1 mW at 550 nm (green-yellow).

CRITICAL ITEMS

FAILURE OF ON-THE-JOB TRAINING/MENTORING

The student received on-the-job training (OJT) on how to exit and leave the lab without crashing the laser (interlock system) as well as how to shut off the laser system if he was the last person to leave the lab. Interviews with staff after the incident found that many thought the student did not have the best set-up or technique. But since he was not a member of the group, no one felt it was their job or responsibility to give him any feedback. The laboratory supervisor, once again, since the student was based outside his group, felt he had little to no supervisory responsibility toward the student. While the student's absentee supervisor felt or believed the local lab supervisor would be watching out for him. Taking all this together, an underexperienced student was left on his own to figure things out, and he turned to just following what he perceived was the correct actions of others.

FAILURE TO MAKE PEOPLE AWARE OF POLICIES

I mention this because the institution had a policy that new students or group members required 30 days of mentoring. Since I was part of an outside investigation team for this incident, I can say I found no evidence that anyone was aware of this policy. It seems as if it was put into the institution's policy manual but no other action was taken to make people aware of the policy.

There's really no need to point out particular actions, other than to say, too many accidents are traced to people doing things that under closer review seem to have no basis in reality or thought.

POSITIVE SIDE OF INCIDENT

This and other safety failures led the institution to reevaluate its approach to safety and polices. A major effort by new management changed the institution's safety culture.

9 A Laser Injury Event at the Idaho National Laboratory
Limitations of Skill-of-the-Craft

Tekla A. Staley

CONTENTS

There has been a long-standing debate within the Department of Energy (DOE) system regarding skill versus process-based approaches to work. After 27+ years at a DOE facility, the discussion still rages on. The craftsman wants to demonstrate his skill and expertise in his trade while management wants a standard, repeatable process no matter who performs the work. So, what is skill-of-the-craft? Idaho National Laboratory (INL)* defines it as:

---

* The INL is a federal site under the jurisdiction of the Department of Energy Idaho Office (DOE-ID), supporting operations activities and research, such as nuclear energy research and development, Department of Homeland Security technologies development and demonstration, and environmental restoration. The INL covers portions of five counties, an area of approximately 890 square miles, in south eastern Idaho. Several active, geographically separated facilities exist throughout the INLs acreage. The information in this chapter is based on ORPS-ID—BEA-SMC-2011-0012 and INL/INT-11-23268. It was originally released under INL/CON-11-23659 and presented at the 2013 International Laser Safety Conference. The statements and opinions in this article are solely those of the author. Nothing in this article represents the work, views, or positions of the Idaho National Laboratory, the Department of Energy, or any of their customers.

Knowledge, skill, and/or physical techniques acquired through education, experience, training (general or specific), or mentoring the performer has acquired over time for a specific discipline or activity.

Work skills are common knowledge to the craft, there is a level of technical proficiency, often through a qualification process, and there are rarely written, detailed, task-specific instructions. Drawings and specifications are common in the field but not the "how-to" type of instruction. A tradesman's knowledge is passed on through verbal and hands-on instruction quite often through apprenticeship programs, and no journeyman or master craftsman teaches alike or passes on the same information. While an expert craftsman is very knowledgeable and skilled in their craft, that knowledge and/or skill is different from person to person; therefore, the steps by which each craftsman gets to the finished product may be very different.

EVENT SUMMARY

A technician at INL received second-degree burns to his left middle and ring fingers while performing maintenance on a Class 4 industrial laser. The task required "target" placement in the beam path to verify mirror alignment. The technician, believing the laser was in a safe mode, reached into the beam pathway to place the target, saw a flash, and immediately withdrew his hand. Investigation included a review of lab-wide safety procedures, equipment-specific operating procedures, maintenance documents, and training records, interviews of maintenance technicians, safety support, and management personnel, and a physical verification of equipment condition. Subsequent weaknesses were identified in training, work direction details, the performance of work following vendor-recommended safe work practices, safety professionals' understanding of a complex system and reliance on the maintenance technician's training, and experience to conduct an effective hazard analysis.

BACKGROUND

Within INL, one project utilizes several Class 4 industrial lasers to fabricate components for various customers. These laser systems require periodic maintenance, including optics cleaning, replacement, and alignment to maintain optimal performance.

The particular laser system involved in this incident, a Cincinnati CL-707 delivery system with a Rofin DC-035 resonator, was installed in 2005. Three basic components comprise this system:

1. The beam generator (resonator), Rofin model DC-035, which contains "slab" electrodes, a rear mirror, an output mirror, a diamond window, a bending mirror, a spherical mirror, a spatial filter, a cylindrical mirror, a power mirror, and shutter (beam dump). In whole, these are referred to as the "internal optics." This portion of the beam is unfocused, with an approximate diameter of ¾", and is available whenever high voltage (HV) is supplied to the "slab" with sufficient energy to produce photons.

2. The beam delivery system, Cincinnati model CL-707, which contains the beam delivery optics external to the resonator cabinet, laser head, gantry, table, Human-Machine Interface (HMI) console and pendant for remote operation. These components are referred to as the "external optics."
3. Material handling system, custom-built by Progressive, which includes vacuum transporters, an overhead gantry, a scrap fork, and a transfer table. This portion of the system has no laser components.

Ten individuals, including electrical control technicians, system engineers, and work planners, have received training from Rofin, Cincinnati, or a combination of the two. Training is specific to the component system. Meaning, Rofin training only covers those components within the resonator cabinet and Cincinnati training only covers the "external optics" portion of the system. The most recent training on the "internal optics" portion of the system occurred in May 2010. Less formal training has taken place when electrical control technicians observed factory representatives performing maintenance on the laser systems while visiting the facility.

Each vendor recommends various preventive maintenance (PM) activities, including a 2000-hour runtime PM. In addition to identifying tasks and providing instructions for maintenance activities, the PM work package also contains a section that identifies hazards and mitigating actions to control arc flash and shock, mechanical motion of the gantry, table and laser head, the high power invisible laser beam, chemical products, pinch points and sharp edges, pressurized gases, noise, and elevated work. Mitigating actions may include lockout and tagout (LOTO), exclusive control of the pendant, barriers and signs, personal protective equipment (PPE), and training. Included in the PM work package are references to vendor procedures that detail maintenance tasks recommended by the vendor. The PM for this particular system was performed using a planned work order (WO), Electrical 2000 Hr. Run Time PMs, and included the following items:

1. Laser cooling system checks.
2. Laser head inspection and cleaning.
3. Ball screw lubrication.
4. Electrical connection checks.
5. Laser gas leak testing.
6. Vacuum testing.
7. Internal optics alignment.
8. RF tube current testing.
9. Mode burns.
10. Safety mat inspections.
11. Gantry material handling system maintenance.
12. Four corner card shots.
13. Beam delivery alignment.

A key component of particular interest in this event is the "pendant." It is a device that allows the electrical control technician to maintain exclusive control of the HMI console when physically separated from the control console. When the pendant is "enabled," control of the laser system is removed from the HMI control console, and is only capable through the pendant control unit, allowing the technicians to perform system functions remotely from areas around the physical footprint of the system and while on either of the work tables.

CHRONOLOGY OF ACTIVITIES

It has previously been established that off-site vendor training was completed approximately 15 months prior to performing this type of PM. Approximately 2 months prior to performing this PM, the WO was updated, including hazard mitigations, reviewed, and approved by the System Engineer, the Maintenance Supervisor, a Safety Engineer, and the Laser Safety Officer (LSO).

In August 2011, about 2 weeks prior to the event on this particular system, nearly identical PM work was performed on a similar laser system without incident.

Beginning on August 29, 2011, PM work on this laser system commenced with a pre-job briefing by the acting Foreman. Three electrical control technicians were assigned to perform the work, and one of them was identified as the person in charge (PIC). The electrical control technicians completed steps 1–24.

The following shift, August 30, 2011, the electrical control technicians continued the PM, at step 25, with a paper shot. This shot is taken downstream of the shutter to determine if alignment within the resonator of the internal optics is needed. The PIC selected the "Mirror Alignment" screen on the HMI and entered/verified the parameters. Control was switched to the pendant, the pendant was enabled, and the "Cycle Start" button was pushed. Once the "Cycle Start" button was pushed, a beam was present, or active, through the beam-forming telescope and z-box (Figure 9.1) to the shutter/beam dump. Thermal paper was positioned at an aperture just beyond the shutter. The PIC used the pendant to flash the shutter and take the shot and an image was burned onto the thermal paper. The image shown on the thermal paper indicated that the internal optics required some attention and possibly alignment.

The system was placed under LOTO to continue with step 26, inspection of the spatial filter assembly and beam absorber for coating damages. The z-box (Figure 9.2) cover was removed to inspect the cylindrical, spherical, and power mirrors. Burn spots were found on the cylindrical and power mirrors, and they were replaced with new mirrors. The spherical mirror was cleaned and reinstalled.

The LOTO was released. Power was restored to the laser system, and the laser head was homed. The PIC, then, selected the "Mirror Alignment" program from the HMI screen, enabled the pendant and pressed the "Cycle Start" button.

Each of the electrical control technicians believed the laser system to be in "simmer" mode. In this mode, HV is on, power is sufficient to create a limited amount of optical radiation in the optical cavity; however, energy is insufficient for optical radiation to exit the diamond window and create a beam throughout the beam-forming telescope and z-box (Figure 9.1).

At this point, indicators were present which, if not overlooked, would have alerted the PIC to the possibility of an exposed beam throughout the beam-forming telescope and z-box (Figure 9.2). The HMI console screen showed a power level of 2500 Watts and the phrase "beam on." Visual indication, through sight glasses on the optical

Slab Laser Beam Formation

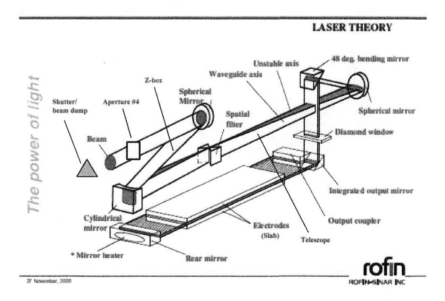

FIGURE 9.1 A schematic of the slab and internal optics.

FIGURE 9.2 Z-box.

cavity (Figure 9.3), showed a bright optical glow. "Beam on" was assumed to be equivalent to "simmer."

After accessing the resonator platform and discussing where to take crosshair shots to verify or adjust the beam alignment, the technicians decided to start with the #4 aperture target (Figure 9.1).

FIGURE 9.3   Optical cavity houses the slab and is equipped with sight glasses.

As one of the three technicians reached into the beam path to insert a crosshair target at aperture #4, they all saw a flash, and the technician immediately removed his hand with the crosshair target. The technician's fingers were burned.

Work associated with the PM was promptly discontinued. The PIC turned off the HV at the pendant, all electrical control technicians left the work, the Forman and Supervisor were notified, and the injured technician was escorted to the on-site medical dispensary for evaluation.

Prior to the end of the shift, a fact-finding meeting was held after which the laser system was secured and placed in a safe configuration for future evaluation.

INVESTIGATION AND ANALYSIS

A cause analyst was charged with conducting an investigation to determine the cause or causes of the event as they relate to the 5 core functions of the Integrated Safety Management System (ISMS). In addition to reviewing training records and work control documentation, conducting interviews with the electrical control technicians, the Foreman, the Supervisor, the System Engineer, and the LSO, the cause analyst utilized two analysis techniques: an Event and Causal Factors Analysis and Barrier Analysis.

Core Function 1—Define the Scope of Work. *Missions are translated into work, expectations are set, tasks are identified and prioritized, and resources are allocated.*

The work scope as identified in the WO was to perform the Electrical 2000 Hr. Run Time PMs as suggested by the vendor. This scope was well defined and understood by the electrical control technicians. They had previously performed the same PM tasks on a similar laser system and had a clear understanding of what was required to complete the work. This core function was met.

Core Function 2—Analyze the Hazards. *Hazards associated with the work are identified, analyzed, and categorized.*

A significant effort was made during 2010 to evaluate the hazards associated with laser maintenance activities. This effort reviewed the existing WO, observed laser maintenance tasks ongoing at the time, and reviewed LOTO practices. The LSO, System Engineer and Safety Engineer were extensively involved with this evaluation.

The hazard evaluation focused on how and for what tasks the LOTO process was incorporated, as some tasks, particularly alignment tasks, must be performed with power available to maintain temperature equilibrium within the optics. While the obvious hazards were identified, the subtleties of how and at what point in the process these hazards would introduce themselves were not well identified. This related back to the programming screens on the HMI console and how the program controls the beam, particularly in "simmer" mode. This core function was not fully met.

Core Function 3—Develop and Implement Hazard Controls. *Applicable standards and requirements are identified and agreed upon, controls to prevent/mitigate hazards are identified, the safety envelope is established, and controls are implemented.*

The details of the task were not fully defined during the work planning phase of this activity. As a result, specific information regarding the development of controls was not well defined. The controls implemented in the WO were based on LOTO to eliminate electrical, mechanical, and beam hazards. Some tasks were never observed or discussed, and, as a result, specific controls for live beam work were not recognized or implemented. Some of these controls, such as turning off the HV versus effecting a full LOTO and checking the beam path with thermal paper prior to placing hands in the beam path were identified in the vendor's training and/or procedures; however, they were not formally incorporated into the WO. Specific controls for live beam tasks that needed a HV switch turned off, as opposed to being placed under LOTO, were not provided in sufficient detail in the WO. There was excessive reliance on the skill, training, and experience of the electrical control technicians to perform the tasks safely. This core function was not fully met.

Core Function 4—Perform Work Within Controls. *Readiness is confirmed and work is performed safely.*

The WO contains a good, but incomplete set of hazards and controls. Hazard controls for mirror replacement (LOTO), open beam paths (establishment of a temporary laser control area (TLCA)), and mechanical motion (LOTO or exclusive control of the pendant) were followed. Other hazard controls requiring lower beam power settings, keeping body parts clear of the beam path, and wearing laser eye protection (LEP) were not followed. Administrative controls such as following safety training tenets and using reference materials from the vendor were not followed. Most of the vendor PM procedures were referenced in the WO, yet the electrical control technicians sought to perform a number of the tasks from memory rather than having the documented procedures physically available for reference and review. This core function was not fully met.

Core Function 5—Provide Feedback and Continuous Improvement. *Feedback information on the adequacy of controls is gathered, opportunities for improving the definition and planning of work are identified and implemented, line and independent oversight is conducted, and, if necessary, regulatory enforcement actions occur.*

Prior to this event, the controls appeared to be adequate. The electrical control technicians had performed the same task a few weeks prior on a different, but similar, laser system without incident. Maintenance management was unaware of any concerns or difficulty following the WO. Safety and the LSO were not aware of any difficulty maintaining compliance with the controls or that the controls were inadequate. This core function was met.

CONCLUSIONS

The direct cause of this event was the misunderstanding regarding beam functions while in the "Mirror Alignment" program screen. Factors contributing to the event include training, work planning, hazard analysis, and work performance issues. The electrical control technicians received relevant training and had sufficient experience to perform the work safely but did not follow some key safe work practices, such as turning off the HV switch and checking for a beam prior to placing a target in the beam path.

The Safety Engineer, a Certified Safety Professional (CSP), had 23 years of experience at the time with extensive knowledge in hazard analysis, LOTO practices, and machine guarding. The LSO, a Certified Industrial Hygienist, CSP, and Certified LSO, has a bachelor's degree in Chemistry and had 27 years of experience at the time with extensive knowledge in physical and health hazard analysis, laser safety, exposure assessment, and LOTO. The System Engineer, with a bachelor's degree in electrical engineering and more than 10 years of experience with laser systems, was new to the facility (fewer than 10 months at the time) and still learning about the laser systems to which he was assigned. Classroom education and field experience in safety, industrial hygiene, and engineering disciplines is no substitute for application-based, hands-on knowledge, rather, it should be complementary to that understanding. It is imperative that craft personnel, with hands-on knowledge and experience, and health and safety and engineering professionals work together to coach each other, provide a clear understanding of the tasks to be performed, expand awareness of the hazards, and provide a thorough hazard analysis.

The WO, while more detailed than most work instructions found in private industry, lacked critical instructions regarding the alignment task and did not reference the specific Rofin procedure for the alignment task. Profound differences in how LOTO requirements are implemented in private industry versus a Department of Energy facility contributed to the limited formal use of vendor procedures and resulted in a heavy reliance on skill-of-the-craft and the performance of tasks from memory. There is a strong possibility that the step-by-step use of the vendor procedure would have prevented the injury-inducing interaction with an active laser beam. The concerns regarding the inconsistency in implementation of LOTO requirements were not discussed or communicated to the LSO or management in the maintenance organization and common ground was not established with the safety engineer, thus, the hazard analysis gap was not identified and there was little opportunity to formally incorporate the vendor procedures such that they complied with in-house LOTO and safety policies.

TRAINING ISSUES

Vendor training on the resonator system was conducted using a Rofin HMI that was not the same as the Cincinnati HMI available to the electrical control technicians at the facility. The technicians were unfamiliar with the differences in system status of the program screen they used. The "Mirror Alignment" screen flashes the shutter (Figure 9.1) and is primarily designed for alignment of the external optics in the Cincinnati portion of the laser system. The "Resonator Service" screen flashes the resonator and is designed for alignment of the internal optics in the Rofin portion of the laser system. The electrical control technicians were more comfortable accessing the "mirror alignment" screen. Although alignment of the internal optics may be achieved using this program, from a safe work practice perspective it is not the preferred method.

The electrical control technicians performing the PM misunderstood the concept of "simmer" mode. As was previously explained, insufficient energy is present to produce a beam beyond the diamond window. Theoretically, no exposed beam would be present, and, thus, no beam hazard. This belief appears to have been created through a combination of circumstances. First, when the technicians had performed crosshair shots in the resonator chamber on the internal optics (Rofin portion of the laser system upstream of the shutter), the "Resonator Service" program was used creating the "simmer" mode. The beam was not present prior to and after taking a shot, as evidence by the paper inserted in the crosshair target. The beam fired (resonator flashed) and returned to "simmer" mode. The Rofin training and procedures require the HV to be switched off prior to placing anything in the beam path. Second, when working on the external optics (Cincinnati portion of the laser system downstream of the shutter), the shutter mechanism prevents the beam from entering the beam delivery system, a level of protection preventing an active beam presence in the external optics beam path. Cincinnati technicians would have used the "Mirror Alignment" program, and pressed the "cycle start" button to generate a beam beyond the diamond window, through the telescope and z-box, to the shutter/beam dump. When the beam fired, the shutter would flash, allowing the beam to travel into the beam delivery system and through the external optics. The "simmer" mode was misconceived as applying to both situations and not corrected by either vendor.

Although electrical and laser operator training and qualifications were identified and required to perform these tasks in the WO, the specialized vendor training courses, retraining requirements every 18–24 months, and methods to track the off-site training were not defined. The difference between the Rofin and Cincinnati training was not recognized.

Work Planning Issues

1. The hazard identification process relied on the skill, training and experience of the electrical control technicians and did not identify some important safety rules established by the vendors. As such, the work order lacked specific controls required to safely perform the work tasks.
2. The WO did not provide a thorough set of work instructions, specific hazard controls, or references to specific vendor procedures to perform the tasks safely. The WO relied on specialized training rather than establishing consistent instructions for the electrical control technicians to follow.
3. Laser eye protection was specified in the WO, and polycarbonate safety glasses were worn. The eyewear worn met ANSI Z87 requirements for safety eyewear; however, the eyewear had not been formally evaluated for laser use and was not labeled with the optical density (OD) and wavelengths for which protection is afforded. It was later determined that safety glasses with polycarbonate lenses were sufficient protection for the tasks to be performed.

4. Applicable vendor procedures were not appropriately incorporated or referenced in the WO.

5. Section I, step 26, of the WO instructions did not provide adequately detailed instruction, reference the correct vendor procedure or discuss applicable safety requirements to perform the task safely. This step instructed the electrical control technician to inspect the spatial filter assembly and beam absorber for coating damages using Rofin procedure VA-19-01-24. The first step in this procedure for ensuring proper spatial filter adjustment is to ensure there is a good alignment from the diamond window through the whole beam forming telescope, which is performed using Rofin procedure VA-19-01-033. This instruction was not included or detailed in the WO, nor was the proper Rofin procedure for that alignment activity referenced or included.

6. The practice of inserting a piece of thermal paper into the beam path to verify the absence of a beam was not formally incorporated into the vendor procedures but was considered skill-of-the-craft knowledge expected of laser technicians. No further reminders were incorporated into the WO, also assuming this was a skill-of-the-craft knowledge for our electrical control technicians.

Work Performance Issues

The electrical control technicians performed portions of the task from memory rather than refer to the Rofin vendor manual procedure VA-19-01-33. During the post-incident investigation interviews, electrical control technicians identified concerns regarding inconsistencies in the implementation of LOTO policies at the facility versus private industry and how the vendor procedures could be executed when those procedures directed the technicians to turn off the HV rather than place the system under LOTO. These concerns were not communicated to the LSO or anyone in the maintenance management chain of command. A lengthy discussion took place with the safety engineer but common ground was not reached regarding when, or if, LOTO was appropriate for all tasks. Because Rofin procedure VA-19-01-33 did not meet the policy expectations for LOTO at the facility, the document was not incorporated or referenced in the WO. The technicians attempted to perform the steps of this procedure from memory and failed to perform an initial key safety step: turn off the HV prior to entering the beam path.

Vendor training encouraged the practice of inserting a piece of thermal paper into the beam path to verify the beam was not present prior to placing anything in the beam path; however, this practice was not consistently practiced by every vendor technician. This variance in practice by the vendor technicians, along with the expectation that an exposed beam was not present in "simmer" mode, reinforced a bad habit of not checking for an active beam prior to interacting with the beam path.

Finally, the electrical control technicians developed a level of confidence based on their experience with other laser systems at the facility. In some cases, this confidence led to a departure from safe work practices encouraged at the vendor training.

Hazard Analysis Issues

During the hazard assessment, the support team (Safety Engineer, LSO, and System Engineer) relied on the electrical control technicians' skills, training, and experience to educate them on the work tasks and activities in order to develop the hazards and controls. After the WO was issued, these support personnel provided insufficient follow-up of the work activities to verify the effectiveness of controls and that the actions taken by the electrical control technicians were appropriate for the level of risk. It is questionable whether the safety engineer or LSO would have recognized some of the hazardous situations given that they do not have the opportunity for, and are discouraged or forbidden from, hands-on interaction with the equipment. These professionals basically oversee the activities at a distance, and, therefore, would not be inside the work envelope to monitor the HMI console or view the state of the optical cavity through the sight glass. The system engineer should have much more detailed knowledge of how the system works and the output parameters of the laser as it operates in different modes, although this is only true for initial commissioning of a system. A direct reflection of our "hands-off" policy is that proficiency degrades over time without periodic practice. Operations and maintenance technicians become the system experts we rely upon with respect to hazard identification and operation.

The assessment of risk relied on expert-based (skill-of-the-craft) controls rather than standards-based controls as was the expectation from line management. When nonstandard controls are used, the effectiveness of those controls must be confirmed. In this case, follow-up monitoring to verify the work was being performed in a safe and compliant manner was not adequately performed. Again, follow-up monitoring by a safety engineer with no laser knowledge and an LSO with no hands-on experience are insufficient to identify the hazards created by subtle differences in different modes of operation.

LESSONS LEARNED

The training audience, content, and frequency of retraining are critical to understanding complex systems, how components from multiple vendors interface, where hidden or subtle hazards may manifest themselves, and for maintaining a level of competency with respect to these complex systems.

These laser systems merge major components from multiple vendors. Each vendor provided training independent of the other and did not recognize the limitations of the training once the components were merged into a single system.

Confidence in one's skill and knowledge sometimes limits critical communication between work groups, allowing assumptions to be followed and insufficient identification of hazards. Work environments which foster open communication between all work groups without hesitation allow more thorough evaluations of work activities.

Providing written work steps of sufficient detail and inclusive of all hazards and controls is always more consistent than performing work steps from memory. When bypassing interlocks and controlling the laser system from the pendant, the following additional controls were implemented:

1. The PIC is solely responsible for operating the pendant and ensuring personnel are clear of hazards.
2. Establish and document conditions under which it is appropriate to switch off the HV versus placing the system under LOTO.
3. Prior to live electrical or beam work, validate the HV status using signal lights and indicators on the HMI console.
4. In addition to noting the shutter position on the HMI console, physically check the status of the shutter position.
5. Prior to entering the beam path, physically verify the absence of a beam with thermal paper.

Although the electrical control technicians, Safety Engineer, LSO, and System Engineer were all well-educated, trained, and skilled in their respective fields, trust and communication are imperative to understanding complex systems, work objectives, and tasks. Objective evaluation and follow-up monitoring are necessary to ensure appropriate safety practices and standards-based controls are implemented to bridge gaps in skill-of-the-craft practices and are utilized to identify when these controls and practices can be improved.

10 Laser Eyewear

Ken Barat

CONTENTS

We can all agree that when most people think of laser safety, laser protective eyewear comes to mind. While there are several elements that go into the function and selection of laser protective eyewear, there are a few items that contribute to laser accidents that center on eyewear.

EYEWEAR AVAILABLE, BUT NOT USED

A fair number of laser accidents can be traced to the user not using laser eyewear when it was available. Is this due to laziness? I would say not—it's rather from problems with the eyewear. Yes, one can misplace their eyewear and not be able to find it or not want to leave the laser use place to get it (say it was left in one's office). More commonly, I believe eyewear it is not used due to dissatisfaction. So, what are the key factors that make folks not want to wear their eyewear?

IT DOES NOT FIT

Like the phrase from a famous legal case, "if it does not fit, you must acquit."* If one's eyewear does not fit, one will not wear it. This excuse with current eyewear is bogus. Any facial shape, even flat nasal bridge, has available eyewear or can be kept on the face with adjustable straps (non-elastic).

I CANNOT SEE WITH MY EYEWEAR ON

Now, this one has some validity. But this is also sometimes driven by price concerns. Many times, a plastic filter will have low visible light transmission, while a higher priced glass or composite filter will give better visualization. A classic example is a green tinted 1064 nm plastic filter and almost clear glass filter for 1064 nm. Eyewear manufacturers are working hard to give each filter greater clarity, and progress has been made with a number of gray-tinted filters for both infrared and visible wavelengths.

Note: Still more exciting is the development of eyewear with cameras built into the frame, just like in a smartphone. Where the lens portion would be opaque and the user will either see a heads-up virtual image or have an image of what is in front of you displayed on one's retina.

COMMON MISTAKES

Judge by Color

Filter color is no indication of a filter's appropriateness for particular wavelengths nor level of optical density (OD) or scale number (LB).

The Higher the Optical Density, the Better

For many filters, the higher the OD, the darker it is, and too much protection can lead to peaking over the eyewear to see what you are doing.

Migrating Eyewear

No one plans on leaving their eyewear on a desk in a different room or at the back end of the room. But that is often where eyewear seems to migrate too. Missing eyewear is like socks in the dryer, nature has a place for them we have not discovered yet. Use eyewear holders.

---

* Johnnie Cochran, *People of the State of California vs. Orenthal James Simpson*, 1995.

Labeling: OD and wavelength fonts are a challenge to read, but must be on the eyewear.

RARE OCCURRENCES: EYEWEAR DEFECTS AND ERRORS

CASE #1: RECEIVING THE WRONG EYEWEAR

An NIST researcher recently received three pairs of laser safety eyewear manufactured by a well-known eyewear manufacturer, that contained filters that did not match what was stated on the product packaging or engraved on the filters themselves. The appropriate polycarbonate filters for the Nd:YAG/Doubled Nd:YAG laser's wavelengths (1064 nm and 532 nm respectively) should have been amber or orange in color (as per catalog image), but the eyewear the researcher received had distinctively blue filters. The researcher noticed the issue when he could see the beam on the bench, which would not have been possible if the filters had been blocking the appropriate wavelengths. NIST metrologists responsible for NIST transmittance calibrations confirmed by direct measurements that the filters received did not meet the specifications engraved on the filters. NIST contacted the manufacturer. The manufacturer confirmed the findings and has put a process in place to prevent the issue of mixing up the filters.

Be aware, check newly received eyewear over, speak up if things do not seem correct or what is expected. A Los Alamos researcher received a set of eyewear with a defect on the coating that allowed laser radiation through unfiltered. These things do happen, but don't let it happen to you.

CASE # 2: A RESEARCHER WEARING EYEWEAR NOTICED
BRIGHT FLASHES ON THE SURFACE OF THE EYEWEAR

On July 18, 2016, a worker at a DOE Lab was aligning a pulsed green laser (527 nm) to a semiconductor wafer when a reflection was directed off the face of the wafer toward the worker's laser protective eyewear (LPE). The worker described the laser beam interaction with the LPE as bright flashes of light across the upper part of the eyewear. The worker was concerned because what he saw (bright flashes) was similar to what is reported by those involved in laser eye injuries. Management was notified and the worker was directed to get an eye examination. No damage was found.

Believing that the LPE may be defective, tests were performed on the eyewear. Good news—it was found to be working as designed. The eyewear did brightly fluoresce when struck by a green laser beam (see Figure 10.1). Other eyewear in the lab was tested, and only the YAG/KTP filter was found to produce the fluorescing phenomenon when struck. The rest produced only a dull spot on the inside of the eyewear during the same test.

What about other manufacturers? Further testing found other filters by different manufacturers resulted in the same fluorescence. Therefore, be aware when using filters used to block visible laser beams, there will be some reemission of light if stuck by a laser beam.

Key item: Anytime one notices any effect on their eyewear, from a flash, flashes, smoke, blurring to composition of the eyewear, step to the side and examine your

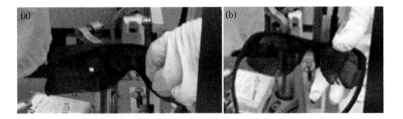

FIGURE 10.1 Filter Fluorescence.

eyewear. It is best to first leave the laser area or block the beam, before removing one's eyewear.

CASE # 3: DIFFUSE REFLECTION

On October 18, 2012, an LANL reported that a worker wearing a pair of Kentek KRZ-C505C laser safety glasses was able to see a diffuse reflection from the 527 nm beam. These glasses are glass lenses with baked-on ceramic coatings that provide an optical density (OD) of 7+ at 532 nm. When the worker changed glasses to a similar pair with the same model lenses, he was no longer able to see the diffuse reflection. A closer look at the incident reveals that the C505C lens provides an OD of only 3.09 at 527 nm. Therefore, why did the second pair block the diffuse reflection? Although they were the same model lens, they were a different shape. It is possible that the difference in shape provided a better solid angle to the reflection.

ULTRAFAST LASER PULSES AND LASER PROTECTIVE EYEWEAR

A study by the U.S. Army Public Health Command demonstrated that the unique properties of ultrashort laser pulses have an effect on laser protective eyewear filters. This effect can lower the optical density of the filters. This work has been confirmed by a study done at NIST in 2016.

DETAILS

Ultrashort pulses can cause shockwaves and bubble formation in the tissues of the eye, which lead to tissue damage.

An injury can occur from an intrabeam exposure or from viewing a reflection of the beam. Most documented injuries from ultrashort laser pulses have occurred when personnel view a reflection of the beam, while eyewear has been removed or one is looking under/over the eyewear. Exposure to just a single pulse, occurring on the order of a trillion times shorter than the blink of an eye, is all that is needed to cause damage.

The unique properties of ultrashort laser pulses make it difficult to protect personnel from them. The lack of protection, in this case from a lower optical density (OD) than that is specified by a manufacturer for such short pulse durations can be contributed to several factors:

- High peak power: Saturable absorption in the filter material may occur which reduces the OD.

- Absorptive material, does not have a quick enough relaxation time.
- Broadband emission: Supercontinuum generation converts a narrow wavelength band to a wide wavelength band which might require a broadband filter.

So how do I protect myself? Remote viewing is the second-best option. Many common laser wavelengths, including those produced by the Ti:sapphire laser commonly used to create ultrashort pulses, can be viewed with inexpensive webcams or digital cameras. Another option is, when feasible, alignments should be made with the laser operating at a reduced energy or a longer pulse width.

The best option for protection during operation would be to enclose the beam path of the laser when emitting ultrashort pulses. This has the added benefit of protecting the optical components from dust.

What about M rated laser protective eyewear? M rated eyewear is narrow notch eyewear made to withstand ultrashort pulses. The problem is the few filters made this way are for very narrow wavelength bands.

So, should I wear my existing eyewear? The answer is YES, but awareness that possible reduction in OD is possible from a direct hit, should make one even more eager to follow good practices.

Material for this section comes from: U.S. Army Public Health Command *FACT SHEET 25-026-0614.*

EYEWEAR SELECTION CRITERIA AND ELEMENTS

Note: The following material is based on the eyewear chapter published in my book, *Laser Safety Tools & Training* (CRC Press, 2008), as well as my work on Z136.1, Z136.7, and Z136.8.

In 1962, Dr. Harold Straub of the U.S. Army Harry Diamond Laboratory, developed the first laser eye protector by installing a 2 x 4-inch, blue-green glass (Schott-type BG–18), filter plate into a standard acetylene welding goggle frame.

The commonest misconception laser users have about laser protective eyewear is that it is the first line of defense against laser radiation. In reality, it should be the last line of defense. Beam containment will do more for a laser user than laser protective eyewear.

If you can eliminate the possibility of eye damage due to enclosing the laser beam path so that NO radiation exposure to the eye is possible, then do so. While critically important, the implementation of laser protective eyewear is always understood to be the second line of defense.

Laser protective eyewear has a valuable role to play in laser safety and presents many challenges to the user and LSO. The remainder of this chapter will deal with the selection and use of laser protective eyewear.

Laser protective eyewear comes in two flavors: full attenuation and alignment eyewear. By full attenuation, I mean this eyewear will completely block the transmission of a direct exposure laser beam from penetrating the eyewear. Conversely, alignment or partial attenuation allows an individual, while wearing laser eyewear, to have some visibility which means some of the beam's energy will pass through the laser protective eyewear.

FIGURE 10.2   Example of eyewear styles.

Frequently, one encounters cases where an LSO recommends, and researchers are then supplied with full attenuation laser eyewear which subsequently is "underutilized" due to research conditions where partial attenuation is required for proper execution of laser related applications (Figure 10.2).

In order to talk about laser protective eyewear, one really needs to understand two terms: Optical Density and Maximum Permissible Exposure. Optical Density or OD is a filtration factor and Maximum Permissible Exposure (MPE) is like the speed limit. Your eye can accept irradiance up to and including the MPE without damage. The higher the exposure or irradiance over the MPE, the greater the damage until a threshold is reached (one you do not want to reach), where the damage itself moderates the energy and damage.

Optical density: OD is a parameter for specifying the attenuation afforded by a transmitting medium. OD is in log units; therefore, goggles with a transmission of 0.000001% can be described as having an OD of 8.0. OD is a logarithmic expression and is described by the following:

$$OD = \log10\left(\frac{Mi}{Mt}\right) \tag{10.1}$$

Where: Mi is the power of the incident beam and Mt is the power of the transmitted beam.

Thus, a filter that attenuates a beam by a factor of 1,000 or 103 has an OD of 3, and one that attenuates a beam by 1,000,000 or 106 has an OD of 6. The OD of two highly absorbing filters stacked together is essentially the sum of two individual ODs. When optical aids are not used, the following relationship may be used when radiant exposure (H) and irradiance (E) are averaged over the limiting aperture for classification:

$$ODreq = log10 \text{ (E or H)/MPE} \tag{10.2}$$

Where: the radiant exposure (E) or irradiance (H) is divided by the MPE. (3) When the entire beam could enter a person's eye, with or without optical aids, the following relationship is used:

$$ODreq = log10[0 \text{ or } Q0/AEL] \tag{10.3}$$

Where: AEL is the accessible emission limit (that is, the MPE multiplied by the area of the limiting aperture) and 0 and Q0 are the radiant power or energy, respectively.

FULL ATTENUATION

Without exception, for Class 4 lasers and Class 3B lasers (when the Maximum Permissible Exposure (MPE) limit is exceeded) it is recommended to provide full attenuation laser protective eyewear in: all UV (nominal 190–380 nm), ocular focus near IR nonvisible (nominal 700–1400 nm) wavelengths, as well as mid to far IR regions. The logic in doing so is quite simple—if one cannot see the beams and they exceed the MPE limits, then there is no reason to do anything other than fully attenuate those same wavelength regions.

Moreover, in the visible regime (nominal 400–700 nm) when detection of the termination point of the visible laser wavelength is NOT required for one's application, then full attenuation of these same visible wavelengths is also recommended.

The LSO is tasked with recommending proper eyewear selection for the wavelength or wavelength region in question to meet the required optical density (OD) for each laser application(s).

Once the small source intrabeam OD for each laser wavelength or wavelength region has been posted, various other ancillary conditions emerge which may both positively (or negatively) impact the intended use of the chosen laser protective eyewear.

To state the obvious: to be effective, laser eyewear MUST be worn. As readily apparent and obvious as that comment is, the single most prevalent cause—by far—of all laser related eye injuries is the fact that laser protective eyewear, while typically available and appropriate to the prevailing laser application, was not worn.

Why? This is where many of the ancillary features such as weight considerations between glass and polycarbonate lenses, "acceptable" versus "unacceptable" visual luminous transmittance (VLT), subjective preferences of comfort and fit, prescription lenses (Rx) capability, propensity of eyewear to fogging, peripheral visual capacity or lack thereof, and so on, come into play.

VISUAL LIGHT TRANSMISSION (VLT)

Undoubtedly, visual light transmission (VLT) and fit are the two most compelling feature in the usage or aversion to usage of laser eyewear. Simply stated, VLT is the mean average percentage of the entire visible spectrum, as weighted for blue spectral responsiveness, which is NOT being filtered by these same lenses. Repeatedly, experience has indicated that the higher the VLT, the higher the likelihood of eyewear usage and consequently laser eyewear safety compliance.

In many research and academic circumstances, overhead room lights may be turned off for a variety of conditions (beam collimation, alignment, and so on) and VLT in these circumstances becomes of preeminent concern. Moreover, laser related electrical hazards, which have caused serious injuries and may include death, must be fully considered in light of diminished visual acuity due to a loss of VLT when wearing laser protective eyewear. Lest we forget, while laser radiation can blind you, electricity can kill you.

Additionally, the distinction between OD and VLT, especially in full attenuation conditions, are sometimes misunderstood or misrepresented. Assumptions abound that a higher OD *necessarily* implies a reduction of VLT. However, reduction of VLT is directly correlated to a higher OD only when visually limiting optical densities are directly attributable to the visible (nominal 400–700 nm) region only.

In laser eyewear attenuation conditions in UV, near, mid, and far IR regions, or a multiwavelength combination thereof, there are relative instances where one may encounter eyewear which possess: high OD, low VLT; high OD, high VLT; low OD, low VLT; low OD, high VLT. In my estimation, any eyewear possessing a VLT at less than 15%–20% is dangerously close to creating a loss of visual acuity where other potential (notably electrical) dangers become considerably more likely.

Therefore, in seeking full attenuation—laser eyewear with appropriate OD values for ones' application(s)—increasing VLT may require certain trade-offs. Typically, this is the decision juncture at which one considers the use of plastic versus glass lenses.

Polycarbonate lenses are lighter in weight than glass lenses. As such, polycarbonate lenses have inherent (and perfectly logical) preference for the user; especially in conditions where protracted usage is required. There are certain common and very prevalent wavelength regions (notably Nd:YAG @ 1064 nm) where glass lenses have higher VLT than polycarbonate lenses. In this instance, the trade-off is of course that while one is increasing the VLT, they are simultaneously increasing the weight of the eyewear and thereby potentially diminishing the perceived comfort of the eyewear.

Fortunately, various manufacturers of both glass and polycarbonate eyewear have noted the general preference for polycarbonate eyewear and have made significant strides in increasing their products' VLT in near IR and certain other visible wavelength regions.

Much of this improved VLT is due to eyewear that obtains all or some of its OD through the reflection of the laser beam. Yes, sufficient energy can be reflected off the eyewear to injure someone else if conditions are right.

COMFORT AND FIT

Comfort and fit considerations are wholly subjective and depend entirely upon individual preferences that each wearer maintains concerning how a specific set of laser protective eyewear feels when worn. Comfort and fit primarily center upon personal preference issues such as: overall comfort when evaluated in terms of short, moderate, or protracted wearing times.

Overall, if a pair of protective eyewear does not fit properly, not only can it not perform its function to the required specifications, but likelihood of it being used decreases. This is true for a respirator, facemask, or laser protective eyewear.

Users want their eyewear to be as natural an extension of their faces as possible. They do not want to be constantly reminded they are wearing eyewear by it being too loose, too tight, too heavy, fogging up, slipping, or other well-known problems.

Therefore, effort placed in finding proper fitting eyewear is well worth the time. One size does not fit all. One solution may be to place a strap across the back to keep the as frame tight as necessary. Another solution maybe flip-down on one's own glasses. Manufacturers offer a range of options in sizes, including new eyewear for slim faces, and for very large faces. There are options for fitting different nasal profiles, including flat or low nasal profiles, and combinations for small faces with flat nasal profiles. Adjustable temple lengths are also helpful, as well as temples with gripping ends. Bayonet temples (the straighter temple) also help in fitting large faces. Choices of laser protective eyewear have come a long way. All users should be able to find their correct pair.

DAMAGE THRESHOLD CONSIDERATIONS

Once one does find the appropriate eyewear with adequate OD to achieve full attenuation and "suitable" VLT, there is yet another trade-off hurdle to ponder, namely damage threshold considerations. As a general "rule of thumb," polycarbonate eyewear can withstand approximately $100 \ W/cm^2$ of direct incident laser radiation for approximately 10 seconds duration prior to "damaging effects" noted on the lenses. Conversely, glass eyewear can withstand approximately 10 times ($\sim1000 \ W/cm^2$) the value of polycarbonate laser eyewear for the same time duration.

With the assumption that a collimated, focused beam is impinging upon a discrete, non-wavering point on the polycarbonate or glass lens, polycarbonate lenses are prone to exhibiting sequentially: a superheated plasma effect at the surface of the lens, degradation of the absorptive dyes (with possible carbonization and darkening effects noted), the emission of smoke, possible noxious odors, the emission of flame, and potential ultimate penetration of the lenses. Conversely, glass lenses are prone to "catastrophic" degradation effects where the accumulation of irradiant energy results in loss of integrity with effects noted as: a popping sound when the beam strikes the glass lens with potential "spider vein" crazing and, with sufficient accumulation of energy, a complete shattering of the glass lens.

Generally speaking, these physical effects for both polycarbonate and glass lenses have readily apparent visual and auditory correlates that forewarn the wearer of an impending damage threshold danger. They do, however, come into consideration when one is deciding upon which trade-offs to implement in order to optimize the likelihood of eyewear suitability and will also be discussed when ultrafast pulse considerations are presented later in the chapter.

SIDE SHIELDS

The ANSI Z136.1 standard, in "Factors in Selecting Appropriate Eyewear" mandates one to "consider" side shields. Overall, the presence of side shields is not an issue that can be "considered" and then decided against. Rather, even though they may impair peripheral vision, I am of a mind that the presence of side shields is mandatory and be commensurate with the level(s) of optical density that the main viewing lenses provide.

The ANSI Z136.1 standard *Safe Use of Lasers* does not require laser protective eyewear to be ANSI Z87 compliant. ANSI Z87 is the standard for safety eyewear; the most common element is impact resistance. Therefore, in evaluating one's eyewear, the question of impact resistance needs to be addressed. Simply, is it needed or not? If not, no further action is needed; if the LSO hazard evaluation is yes, it is required, then one has 3 choices:

1. Obtain a pair of laser eyewear that is compliant with Z87 (most polymer eyewear are compliant).
2. Wear safety glasses over the laser eyewear.
3. Have glass laser eyewear hardened to meet Z87.

Choice 2 can affect comfort or the ease of wearing the protective eyewear and general vision, while Choice 3 will affect the cost of the eyewear.

PRESCRIPTIONS

There are several options for prescription eyewear. These include eyewear with prescriptions ground into the glass laser lens, eyewear which holds prescription inserts, and eyewear with flips—polymer prescriptions in the base or the flip. For ground lenses, the frame selections have widened to include titanium frames and frames with adjustable temples (Figure 10.3a,b).

WEIGHT

Weight of eyewear is a particular concern in the consideration of acquiring multiwavelength or prescription eyewear. Depending on wavelength combination, 7 mm of glass is not unheard of. This thickness of glass 2–3 times a normal prescription eyewear may prove to uncomfortable for a user to wear for extended periods. This can lead to a lack of productivity or times of no eye protection. Some breakthrough in polycarbonate prescription flips and over-glasses may help improve this item.

LABELING

ANSI Z136.1 and IEC require laser protective eyewear to be labeled with the wavelength and OD it is intended for. The laser eyewear manufacturer will imprint on the eyewear the most common range of wavelengths and OD for a particular pair. For the vast number of laser users, this is satisfactory. Always remember the guarantee of protection is only made for the wavelengths imprinted on the eyewear frame. Even curves for the eyewear are just a generalization. Unless you know the lot number and have curves for that run, only the imprinted OD is guaranteed. A small segment of users are using the eyewear for wavelengths not listed on them. Curves

(a)

(b)

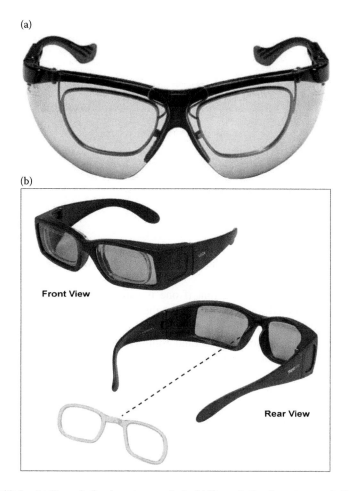

Front View

Rear View

FIGURE 10.3 (a) Prescription insert example 1, (b) Prescription insert example 2.

and other documentation provided by the eyewear manufacturer or distributor will show the OD at the desired wavelength. To be compliant, the facility LSO will have to label the eyewear or post the information where the eyewear is stored and have a way to identify which pair is which (Figure 10.4).

ULTRAFAST LASERS

Testing by the Army branch at Brooks Air Force base has shown a nonuniform bleaching effect on standard laser eyewear against ultrafast pulses. This relates back to the relaxation time of the absorption molecules. Not all eyewear for ultrafast pulses demonstrate this effect, but a significant number do to make it a real concern. Therefore, for ultrafast laser users who wish for full protection, they will need to check with the manufacturer of the eyewear for their testing results to verify suitability of the eyewear for their use. Usually, the manufacturer can provide a sample piece of the lens

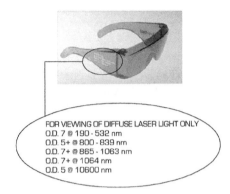

FOR VIEWING OF DIFFUSE LASER LIGHT ONLY
O.D. 7 @ 190 - 532 nm
O.D. 5+ @ 800 - 839 nm
O.D. 7+ @ 865 - 1063 nm
O.D. 7+ @ 1064 nm
O.D. 5 @ 10600 nm

FIGURE 10.4   Eyewear labeling.

for testing with a power meter in the actual application, to verify the appropriateness of the lens in question.

It is imperative to recognize that if one is using ultrafast lasers (particularly regeneratively amplified sources) there exists the potential that optical density values may be compromised should beam exposure to one's laser eyewear occur. Should temporary or permanent loss of OD (and commensurate exposure levels in excess of applicable MPE values) occur as a consequence of these conditions, obvious detrimental eye safety effects become plausible. The core safety issue surrounding laser protective eyewear and femtosecond lasers is as follows: in certain ultrafast (femtosecond) operating conditions, saturable absorption effects with calculable losses in purported optical density values of the femtosecond subjected laser eyewear have been observed.

It is the intention of ANSI committees involved in this matter that the underlying mechanisms of the degradation effects so noted be investigated and, to the greatest extent possible, elucidated for everyone's general understanding.

ADDITIONAL CONSIDERATIONS

Another important consideration is anti-fog capabilities, especially for goggles. Multiwavelength operations have special questions, as the more wavelengths you try to remove with one pair of eyewear, typically, the darker the eyewear gets. You can try flip options or more than one pair to alleviate this problem. Laser inscribed markings (printed ones wash off when cleaned) also help the longevity of the eyewear, as well as UV inhibitors to prevent darkening over time in polymer eyewear. Finally, cost is important, but you must also consider what the cost of an eye is.

PARTIAL ATTENUATION (AKA ALIGNMENT EYEWEAR)

As noted previously, there are frequent occasions in laser-related research applications where investigators need to view the termination point of visible laser sources. These

TABLE 10.1
Recommended Optical Density for Alignment (EN208)

Scale #	OD	Max Instantaneous Power Continuous Wave Laser (W)	Maximum Energy for Pulsed Lasers (J)
R1	1–2	0.01	2×10^{-6}
R2	2–3	0.1	2×10^{-5}
R3	3–4	1.0	2×10^{-4}
R4	4–5	10	2×10^{-3}
R5	5–6	100	2×10^{-2}

same beam alignment conditions are also acknowledged to produce a notable amount of laser-related eye injuries.

The purpose of alignment eyewear is to allow the user visualization of the beam while lowering the intensity of any beam that is transmitted through the user's eyewear to a Class 2 level. To address this issue, there is an existing European Norm which recommends optical density for alignment eyewear versus the output of lasers used (Table 10.1).

Therefore, for "alignment" laser eyewear to be effectively utilized, preferentially all of the following conditions should be in place: administrative liability acknowledgment and acceptance of same, acknowledgment of potential hazards with the utilization of eyewear that does not protect one against small source intrabeam or specularly reflected exposures and finally, collaborative agreement between the LSO and researcher(s) of alignment eyewear safety protocols and appropriate "alignment" laser protective eyewear. Once these preliminary "philosophical" protocols are established, the implementation of alignment eyewear can proceed forward.

The trouble with EN207 is, on the pulse side, it does not cover today's laser pulses nano, pico, and femtoseconds. My experience considers a decrease of 1.4 OD is the maximum for alignment eyewear used for pulsed lasers.

LOW-LEVEL ADVERSE VISUAL EFFECTS

At exposure levels below the maximum permissible exposure (MPE), several adverse visual effects from visible laser exposure may occur. The degree of each visual effect is strongest at night and may not be disturbing in daylight. These visual effects are:

a. *Afterimage.* A reverse contrast, shadow image left in the visual field after a direct exposure to a bright light, such as a photographic flash. Afterimages may persist for several minutes, depending upon the level of adaptation of the eye (that is, the ambient lighting).

b. *Flashblindness.* A temporary visual interference effect that persists after the source of illumination has been removed. This is similar to the effect produced by a photographic flash and can occur at exposure levels below those that cause eye injury. In other words, flashblindness is a severe afterimage.

   c. *Glare.* A reduction or total loss of visibility in the central field of vision, such as that produced by an intense light from oncoming headlights or from a momentary laser pointer exposure. These visual effects last only as long as the light is actually present. Visible laser light can produce glare and can interfere with vision even at exposure levels well below those that produce eye injury.

   d. *Dazzle.* A temporary loss of vision or a temporary reduction in visual acuity.

   e. *Startle.* Refers to an interruption of a critical task due to the unexpected appearance of a bright light, such as a laser beam.

Surprise factors that affect laser eyewear coatings are shown in Figure 10.5.

A common question is: Can I stand in front of any beam? The simple answer is that you should never stand directly in the beam of the beam and the other answer is NO, Stupid. We are now in the realm of laser filter damage threshold (maximum irradiance). At very high-beam irradiances, filter materials that absorb or reflect the laser radiation can be damaged. It, therefore, becomes necessary to consider a damage threshold for the filter. Typical damage thresholds from Q-switched, pulsed laser radiation fall between 10 and 100 J cm^{-2} for absorbing glass, and 1 to 100 J cm^{-2} for plastics and dielectric coatings. Irradiances from CW lasers, which would cause filter damage, are in excess of those that would present a serious fire hazard, and therefore, need not be considered. Personnel should not be permitted in the area of such lasers.

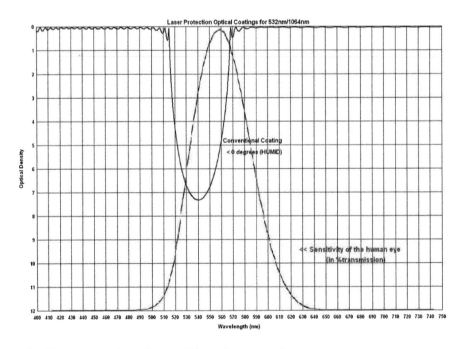

FIGURE 10.5 Humid vs. Dry conditions effect on coatings.

EUROPEAN SYSTEM

A few words about international standards: EN stands for European Norm, in U.S. terms, a standard.

EN 207

Laser eye protection products require direct hit testing and labeling of eye protectors with protection levels, such as D 10600 L5 (where L5 reflects a power density of 100 MegaWatt/m^2 as the damage threshold of the filter and frame during a 10 seconds direct hit test at 10,600 nm). Filter and frame must both fulfill the same requirements. It is not acceptable to select glasses according to Optical Density alone. The safety glasses must be able to withstand a direct hit from the laser for which they have been selected for at least 10 seconds (CW) or 100 pulses (pulsed mode).

EN 208

This norm refers to glasses for laser alignment. They will reduce the actual incident power to the power of a class II laser (<1 mW for continuous wave lasers). Lasers denoted as class II are regarded as eye safe if the blink reflex is working normally. Alignment glasses allow the user to see the beam spot while aligning the laser. This is only possible for visible lasers (according to this norm "visible lasers" are defined as being from 400 to 700 nm). Alignment glasses must also withstand a direct hit from the laser for which they have been selected, for at least 10 seconds (CW) or 100 pulses (pulsed mode).

EN 60825

Requires that laser safety eyewear provide sufficient optical density to reduce the power of a given laser to equal to or less than the listed Maximum Permissible Exposure levels (MPE). It allows specification according to optical densities in extreme situations, but recommends the use of EN 207 with a third-party laser test. In neither standard is a nominal hazard zone allowed; the only consideration is protection against the worst-case situation, such as direct laser radiation.

See Tables 10.2 and 10.3.

TABLE 10.2
Code Definitions

Testing Conditions for Laser Type	Typical Laser Type	Pulse Length (s)	Number of Pulses
D	Continuous wave laser	10	1
I	Pulsed laser	10^{-4} to 10^{-1}	100
R	Q Switch pulsed	10^{-9} to 10^{-7}	100
M	Mode-coupled pulse laser	$>10^{-9}$	100

TABLE 10.3
Scale Numbers

		Power and Energy Density (E, H) for Testing the Protective Effect and Stability to Laser Radiation in the Wavelength Range								
		180 nm to 315 nm			>315 nm to 1400 nm			>1400 nm to 1000 nm		
					For test condition					
		$D > 3\text{-}10$	$I, R\ 10^3$ to $3\text{-}10^4$	$M < 10^{-3}$	$D > 5\text{-}10^{-4}$	$I, R\ 10^9$ to $5\text{-}10^{-4}$	$M < 10^{-9}$	$D > 0,1$	$I, R\ 10^3$ to $0,1$	$M < 10^{-9}$
	Maximum Spectral Transmittance for Laser Wavelength									
Scale Number	$\tau(\lambda)$	$E_D\ W/m^2$	$H_{I,R}\ 3/m^2$	$E_M\ W/m^2$	$E_D\ W/m^2$	$H_{I,R}\ 3/m^2$	$H_M\ 3/m^2$	$E_D\ W/m^2$	$H_{I,R}\ 3/m^2$	$E_M\ W/m^2$
L1	10^{-1}	0.01	$3\text{-}10^2$	$3\text{-}10^{11}$	10^2	0.05	$1,5\text{-}10^3$	10^4	10^3	10^{12}
L2	10^{-2}	0.1	$3\text{-}10^3$	$3\text{-}10^{12}$	10^3	0.5	$1,5\text{-}10^2$	10^5	10^4	10^{13}
L3	10^{-3}	1	$3\text{-}10^4$	$3\text{-}10^{13}$	10^4	5	0.15	10^6	10^5	10^{14}
L4	10^{-4}	10	$3\text{-}10^5$	$3\text{-}10^{14}$	10^5	50	1.5	10^7	10^6	10^{15}
L5	10^{-5}	100	$3\text{-}10^6$	$3\text{-}10^{15}$	10^6	5.10^2	15	10^8	10^7	10^{16}
L6	10^{-6}	10^3	$3\text{-}10^7$	$3\text{-}10^{16}$	10^7	$5\text{-}10^3$	$1,5\text{-}10^2$	10^9	10^8	10^{17}
L7	10^{-7}	10^4	$3\text{-}10^8$	$3\text{-}10^{17}$	10^8	$5\text{-}10^4$	$1,5\text{-}10^3$	10^{10}	10^9	10^{18}
L8	10^{-8}	10^5	$3\text{-}10^9$	$3\text{-}10^{18}$	10^9	$5\text{-}10^5$	$1,5\text{-}10^4$	10^{11}	10^{10}	10^{19}
L9	10^{-9}	10^6	$3\text{-}10^{10}$	$3\text{-}10^{19}$	10^{10}	$5\text{-}10^6$	$1,5\text{-}10^5$	10^{12}	10^{11}	10^{20}
L10	10^{-10}	10^7	$3\text{-}10^{11}$	$3\text{-}10^{20}$	10^{11}	$5\text{-}10^7$	$1,5\text{-}10^6$	10^{13}	10^{12}	10^{21}

ADDITIONAL REFERENCES

Lyon, T.L., and Marshall, W.J. Nonlinear Properties of Optical Filters—Implications for Laser Safety. *Health Phys.* 51, no. 1 (1986):95–96.

Mclin, L. A Case Study of a Bilateral Femtosecond Laser Injury, *Proceeding of the International Laser Safety Conference 2013*, Laser Institute of America, paper #904.

Rockwell, B., Thomas, R., and Vogel A. Ultrashort Laser Pulse Retinal Damage Mechanisms and their Impact on Thresholds, *Medical Laser Application*, 25 (2010): 84–92.

Stolarski, J., Hayes K.L., Thomas R.J., Noojin, G.D., Stolarski, D.J., and Rockwell, B.A. Laser Eye Protection Bleaching with Femtosecond Exposure. *Proc. SPIE 4953, Laser and Noncoherent Light Ocular Effects: Epidemiology, Prevention, and Treatment III* 4953 (2003):177–184.

11 Risk Assessment

Randy Paura

CONTENTS

Continuous improvement is a mantra for a business to achieve operational efficiency, where various forms of a cost-benefit analysis are on-going for all facets of its activities. Occupational health and safety is also fundamental to an organization's livelihood and success. Inherent hazards must be known, their potential risks evaluated and prioritized for elimination or mitigation with safety control measures, and reassessed to determine that any residual risk of a hazard is as low as reasonably practicable (acceptable). The core of this structured process is known as risk assessment and its implementation is known as risk management.

An organization's resources, like opportunities, are valuable: they are not to be wasted. Risk assessment allows an organization to efficiently execute its responsibility for the general duty of care to its employees and affected personnel.

When it comes to occupational health and safety with lasers, risk assessment can have many forms in terms of its documentation, ranging from qualitative to quantitative, although the principles remain the same.

This chapter provides context for the LSO involved with leading a risk assessment for a laser-based project or as part of a team on a project involving lasers. Summary guidance is provided on risk assessment terms, principles, and format.

BACKGROUND

For traditional laser systems, ensuring conformance to regulatory build standards and validating safe use in accordance to ANSI Z136 will achieve functional safety and fulfill the general duty of care obligation by the employer to the operators and employees.

The structure of the ANSI Z136 consensus standard series for the safe use of lasers consists of:

- Determine the hazard class of a laser
- Apply the respective control measures, summarized in Tables 10 and 11 of Z136
- Monitor and audit the safety control measures

Embedded within this mature classification scheme is the risk assessment for the laser's capability of injuring personnel or interfering with task performance. It is important to note that Class 4 lasers start at 0.5 Watts of continuous wave power and there is no subsequent breakpoint for a higher hazard class based upon a "next level" bio-effect threshold. Within Class 4, increasing laser power/energy levels represent an increasing scale of those hazards which require commensurate safety control measures.

Below is a simplified version of a laser hazard classification scheme, referencing both ANSI Z136 and IEC 60825 series (Figure 11.1).

Understanding risk assessment will allow for greater appreciation and utilization of the hazard classification scheme of ANSI Z136 and in generating the appropriate documentation. More importantly, it will enable the suitable depth and scope of a written risk assessment for those lasers and their applications as warranted.

The material presented herein is a synthesis of risk assessment principles and practices for the competent and responsible LSO to ensure that all reasonably foreseeable risks are addressed, ensuring Light is Applied Safely, Efficiently, and Reliably.*

---

* Acronym attributed to Dr. D. Sliney

Hazard Class	Long-term exposure		Short-term (accidental) exposure			
Hazard Class	Eye Magnified	Eye Intra-beam	Eye Magnified	Eye Intra-beam	Eye Diffuse	Skin Exposure
Class 1	Safe	Safe	Safe	Safe	Safe	Safe
Class 1M	At Risk	Safe	At Risk	Safe	Safe	Safe
Class 2	At Risk	At Risk	Safe	Safe	Safe	Safe
Class 2M	At Risk	At Risk	At Risk	Safe	Safe	Safe
Class 3R	At Risk	At Risk	Some Risk	Some Risk	Safe	Safe
Class 3B	At Risk	At Risk	At Risk	At Risk	Some Risk	Some Risk
Class 4	At Risk	At Risk	At Risk	At Risk	At Risk	At Risk

FIGURE 11.1   Laser hazard classification scheme: simplified.

WHAT IS A RISK ASSESSMENT?

Risk assessment is a term used to describe the portion of the risk management process or method. Risk management consists of:

- Identification of hazards and their factors that have the potential to cause harm (hazard identification);
- Analysis and evaluation of the potential for these hazards to be realized, (risk analysis);
- Prioritization of the risks from those needing the greatest attention for safety control measures to the least, (risk evaluation);
- Determination of appropriate ways to eliminate the hazard, or control the risk when the hazard cannot be eliminated (risk control);
- This process is iterative until the subject risk is acceptable.

In many safety circles, including those agencies responsible for workplace safety, risk assessment is used interchangeably for risk management. For the purposes of this chapter, the topic of risk analysis and risk control will also be covered as these are crucial for the risk assessment component of an organization's risk management process. Figure 11.2 shows the relationship of the various components in the risk management process.

Risk management is the structured examination to identify those processes, operations, equipment, situations, environment, and so on, that may cause harm to personnel or property. After identification of potential hazards are made, next is the determination of how severe and likely that hazard may be realized; the product of these factors is referred to as the risk. With the evaluations complete, the hazards, and their respective risks evaluated, will allow for prioritization and reassessment with safety control measures applied until the residual risk(s) are as low as reasonably practicable and acceptable. The objective is to effectively eliminate or control the harm from happening.

FUNDAMENTAL CONCEPTS

Risk and hazard are used interchangeably by some. It is true that these two terms are related, but they have very distinct definitions. A hazard is what can cause harm or

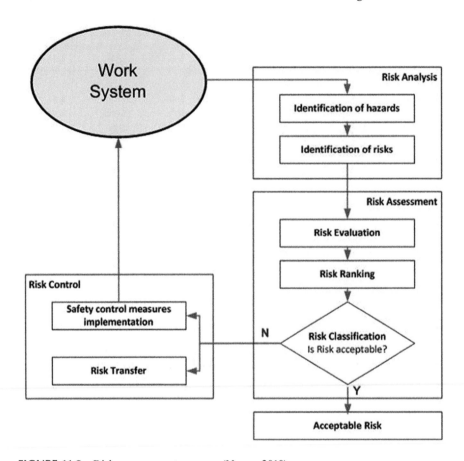

FIGURE 11.2   Risk management process. (Nunes, 2010)

injury to a person or property. There are two major groups of hazards associated with lasers or equipment containing lasers: beam and non-beam hazards.

Risk is the likelihood of an occurrence of a hazardous event or exposure and the severity of injury or ill health that can be caused by the event or exposure. For the technically inclined, at its most basic, risk is a function of two variables:

Risk = f(severity of the hazard, likelihood or probability of occurrence of harm)

Lastly, the concept of acceptable risk needs to be understood. In broad terms, acceptable risk is that which has been reduced to a level that can be tolerated by the organization having regard to its legal obligations and its own OHS policy. A commonly referenced objective for risk is that it be As Low As Reasonably Acceptable (ALARA) or As Low As Reasonably Practicable (ALARP). One reference for determining when a risk is ALARA/ALARP is where additional safety control measures do not bring a corresponding reduction in the risk valuation for a hazard. Just because one has determined that ALARA/ALARP has been achieved, the final question remains, is it acceptable?

Whereas this chapter is intended to provide summary guidance on this topic of risk assessment for lasers, the reader is encouraged to reference the more thorough consensus standard(s) that may apply for their environment and application. Such include but are not limited to:

DEALING WITH PRINCIPLES

- ANSI B11.0-2010, "Safety of Machinery—General Requirements and Risk Assessment"
- ISO 12100:2010, "Safety of machinery—General principles for design—Risk assessment and risk reduction"

Providing Application or Functional Requirements:

- ANSI/RIA R15.06-2012 "American National Standard for Industrial Robots and Robot Systems—Safety Requirements"
- IEC 61508 series "Functional Safety of E/E/PE"
- IEC 62061 "Safety of machinery: Functional safety of electrical, electronic and programmable electronic control systems"
- ISO 13849 series "Safety of machinery—Safety-related parts of control systems"
- MIL STD 882E "USA Department of Defense—Standard Practice—System Safety"

TERMS AND DEFINITIONS

Within the framework of Occupational Health and Safety, the following terms need to be clearly understood:

Hazard: The potential to cause harm—which can include substances or machines, processes, methods of work, or other aspects of an organization.

Risk: The likelihood that the harm from a particular hazard is realized.

Hazard identification: The process of finding, listing, and characterizing hazards.

Risk analysis: A process for comprehending the nature of hazards and determining the level of risk.
- Risk analysis provides a basis for risk evaluation and decisions about risk control.
- Information can include current and historical data, theoretical analysis, informed opinions, and the concerns of stakeholders.
- Risk analysis includes risk estimation.

Risk evaluation: The process of comparing an estimated risk against given risk criteria to determine the significance of the risk. It enables prioritization for the implementation of safety control measures.

Risk control: Actions implementing risk evaluation decisions, which includes safety control measures.

- Risk control can involve monitoring, re-evaluation, and compliance with decisions.

In dealing with likelihood and consequence of a hazardous event (occurrence), it is important to be mindful that different sectors may have different grades, as will be explained following. For now, the most basic of grading is provided below.

Likelihood of occurrence (probability):

- Very likely: Near certain to occur
- Likely: May occur
- Unlikely: Not likely to occur
- Remote: So unlikely as to be near zero

Consequence (severity):

- Catastrophic: Death or permanently disabling injury or illness (unable to return to work). Includes loss of vision in both eyes, loss of multiple limbs, quadriplegic
- Serious: Severe debilitating injury or illness requiring more than first aid (able to return to work at some point). Includes loss of vision in one eye, loss of a limb.
- Moderate: Significant injury or illness requiring more than first aid (able to return to same job)
- Minor: No injury or slight injury requiring no more than first aid (little or no lost work time).

Using these definitions, a risk matrix can be constructed, whereby the hazard potential can be evaluated.

WHY IS RISK ASSESSMENT IMPORTANT?

Risk assessments are very important as they form an integral part of an occupational health and safety management plan. They help to:

- Create awareness of hazards and risk.
- Identify who may be at risk (employees, cleaners, visitors, contractors, the public, and so on).
- Determine whether a control program is required for a particular hazard.
- Determine if existing control measures are adequate or if more should be done.
- Prevent injuries or illnesses, especially when done at the design or planning stage.
- Prioritize hazards and their corresponding safety control measures.
- Meet legal requirements where applicable.
- Demonstrates "due diligence" in meeting the "general duty" clause

Increasingly, an end customer can require a risk assessment from the provider of a laser system, even if it is not a regulatory requirement.

WHERE ARE RISK ASSESSMENTS NOTED/REQUIRED BY REGULATIONS?

It is a mistake to assume that since a regulation does not explicitly spell out a risk assessment for an application/process/equipment, that a risk assessment is not needed or required. Like any safety regulation, what has been codified is not normally the product of foresight, rather it is in response to an accident. Risk assessments, when identified in regulations, are addressing those most prominent hazards that have had a tragic event. Demonstration of due diligence in fulfilling the obligations of the general duty clause is best achieved with a documented risk assessment.

The trend with member states of the International Labour Organization (such as Europe, Canada, Australia, New Zealand) has been to formally identify the principle of risk assessment in their regulations.

Countries that have legislated risk assessment as an "A-Level" standard include:

- Europe (Framework Directive 89/391/EEC)
- United Kingdom (1999 HSE)
- Canada
- Australia
- New Zealand

The above list does not include those countries which accept and recognize risk assessment through interpretive notices, legal precedence, or under the fold of Recognized and Generally Accepted Good Engineering Practices (RAGAGEP).

Within the United States, risk assessments are mandated in OSHA 29 CFR 1910.119: *Process Safety Management of Highly Hazardous Chemicals*, and EPA 40 CFR Part 68: *Risk Management Plan*. Of interest is where OSHA prescribes the principle of ensuring that the merit, suitability, and specification of PPE against a hazard be addressed with a hazard assessment:

1910.132(d)(1)
The employer shall assess the workplace to determine if hazards are present, or are likely to be present, which necessitate the use of personal protective equipment (PPE)...

1910.132(d)(2)
The employer shall verify that the required workplace hazard assessment has been performed through a written certification that identifies the workplace evaluated; the person certifying that the evaluation has been performed; the date(s) of the hazard assessment; and, which identifies the document as a certification of hazard assessment.

Whereas OSHA requires a written certification that the hazard assessment has been performed, it does not prescribe that the hazard assessment itself be in writing. To fulfill the spirit, intent, and letter of the law, a risk assessment would:

- Determine the inherent hazard(s) and their risk(s)
- Assess the effectiveness of safety control measures in place
- Evaluate the residual risk(s) posed to personnel
- Identify if PPE is warranted and enable their specifications and applicability

Whether knowingly or not, employers make a risk assessment regarding the health and safety of their employees and affected personnel when determining the level of protective measure put in place for the operation of equipment and its processes. Due diligence is demonstrated if the employer can present some form of documented risk assessment pertaining to its equipment and processes. Remember, due diligence is what counts before there is an accident or incident.

It is advisable that risk assessment should be done on a periodic basis (such as a yearly safety audit) or when a change is introduced in the workplace, such as with the introduction of new (or alteration of) equipment or procedures. This is encompassed with the laser safety audits noted in the normative portions of ANSI Z136.1:2014 (A1.2)(l); Z136.8:2012 (A2.2)(m); Z136.9:2013 (A1.2)(l).

WHAT IS THE GOAL OF RISK ASSESSMENT?

The goal of the risk assessment process is to eliminate the unknowns affecting workplace safety, through determination of:

- What are the inherent hazards?
- What are their risks?
- How effective are the control measures?
- What are the residual risks to personnel?

The object/goal is to enable knowledge within the workplace of affected personnel and stakeholders of what hazards exist, how they are contained and controlled, and what the residual risks are. An informed workplace will be able to answer the following questions:

a. What can happen and under what circumstances?
b. What are the possible consequences?
c. How likely are the possible consequences to occur?
d. Who is at risk (exposure base/population)?
e. Is the risk controlled effectively, or is further action required?
f. Is the risk As Low As Reasonably Practicable (ALARP)?
g. Is the risk acceptable, per regulations and facility OHS policy?

WHAT IS ACCEPTABLE RISK?

By now, this question should have come to mind, and there is considerable history behind the legal and societal evolution of acceptable risk. This subject alone has seen many attempts to identify, understand, and quantify. The following is provided as baseline guidance for appreciation, as acceptable risk is relative to its environment, subject hazard of interest, and whether the risk is voluntary (such as the driver of a vehicle for transport: a car) or involuntary (such as a passenger in a vehicle: an airplane).

Zero risk is the ideal, altruistic objective for any undertaking with an inherent hazard or potential for failure. In practice, risk is inherent with any human activity or undertaking. Our appreciation for risk is based upon our experiences and perception, hence it can be problematic to define and quantify.

The World Health Organization (WHO) has tackled this concept, when attempting to understand society's definition of what is an acceptable risk when it comes to their health.*

To provide context, in a general sense, a risk is acceptable when:

- It is As Low As Reasonably Acceptable (ALARA):
  - Falls below an arbitrarily defined threshold (e.g., agreed upon by affected stakeholders)
  - Falls below a level already established and accepted (by custom, practice, norm)
  - Falls below a comparative probability threshold that is accepted (e.g., a disease burden in a community/society)
- It is As Low As Reasonably Practicable (ALARP):
  - The cost of reducing the risk would exceed the costs saved
  - The cost of reducing the risk would exceed the costs save when the "costs of suffering" are also accounted for
  - The opportunity costs would be better invested on other, higher priority health problems
- Recognized professionals say it is safe
- The general public is in acceptance, or there is consent by silence (non-objection)
- Governing authorities deem it as acceptance

For the technically inclined, the above is suitable for a very general frame of reference, but quantification provides more concrete means to establish safety goals, and to verify that safety is maintained. The genesis of acceptable risk levels is born out of the nuclear power plant era (regarding societal exposures) and has been continuously studied, organized, debated, ligated, and defined.

When it comes to probability assessments regarding events and occurrences for workplace safety, it should be distinguished that very high value/risk or programs or projects may have detailed numerical analysis. A good example is the Apollo program which quantified reliability and robustness for individual components of its Saturn V launch vehicle, command module, and lunar module for a stated mission success rate, that all stakeholders bought into and strove toward.

For an industrial laser system, employing some purpose-built components, but mostly commercially available components, a detailed numerical risk assessment is not normally practicable. However, knowing the principles behind a risk assessment, it is possible to establish certain safety performance criteria such as the Safety Integrity Level, and Performance Level of the safety related controls elements and safety circuit.

---

* An appropriate framework for understanding this topic can be found in the World Health Organization's "Chapter 10: Acceptable Risk," in *Water Quality: Guidelines, Standards and Health*, edited by Lorna Fewtrell and Jamie Bartram (London: IWA Publishing, 2001). © 2001 World Health Organization.

Timeframes for a risk valuation can be in terms of one's lifetime or per year, or in terms of failure when the demand for the safety function is called upon.

The Health and Safety Executive in the United Kingdom has established the following acceptable risk breakpoints for a person dying in one year (per annum, pa):

- 1 in 1000 (10^{-3}_{pa}) as "just about tolerable risk" for any substantial category of workers for any large part of a working life
- 1 in 10,000 (10^{-4}_{pa}) as the "maximum tolerable risk" for members of the public from any single (non-nuclear) plant
- 1 in 100,000 (10^{-5}_{pa}) as the "maximum tolerable risk" for members of the public from any new nuclear power station
- 1 in 1,000,000 (10^{-6}_{pa}) as the level of "acceptable risk" at which no further improvements in safety need to be made

These probabilities were benchmarked against the risk of 1 in 1,000,000 for being electrocuted at home (10^{-6}_{pa}) (RCEP 1998). Note that the term "one in a million" is familiar: it is a per year risk in the United Kingdom, whereas in the United States, it is a lifetime risk.

Knowing the above societal acceptable risk probabilities, the functional performance of safety critical components can be better appreciated. Safety related controls have strict performance, reliability, robustness, and integrity requirements, with corresponding grade levels. Appreciation for such can be accomplished with the concept of Probability of Failure on demand (PF_d). PF_d can be likened to the probability that a safety element will fail when required to perform its function. A recent study determined that when faced with a panic situation and a process termination is required (via an e-stop), an unskilled operator will make an incorrect decision (fail to initiate the e-stop) 1 out of 10 times ($PF_d = 10^{-1}$). A skilled operator would reduce this risk of failure to 1 in 100 ($PF_d = 10^{-2}$). In contrast, a safety related component must be on the order with a $PF_d < 10^{-6}$ (Table 11.1).

Identifying the performance level of a safety circuit will also drive the corresponding integrity of the safety measures implemented to contain and control a hazard and its risk potential.

The previously noted safety controls performance standards qualify limits for reliability classification by the probability of dangerous failures per hour. Table 11.2

TABLE 11.1
Failure Rate

System	Claimed Failure Rate or Probability of Failure on Demand (PF_d)
Operator action	10^{-1}/demand (typical)
	10^{-2}/demand (best)
Non-safety related controls system (e.g., PLC)	Not better than 10^{-5}/hr
High integrity protective system (e.g., Safety PLC)	Not better than 10^{-9}/hr

---

TABLE 11.2

Safety Controls Performance Standards Qualify Limits

ISO 13849 Performance Level (Pl)	IEC 62061 Safety Integrity Level (SIL)	IEC 61508 Safety Integrity Level (SIL)	PFH$_D$ (Probability of Dangerous Failure per Hour)
a	None	None	$\geq 10^{-5}$–$<10^{-4}$
b	1	1	$\geq 3 \times 10^{-6}$–$<10^{-5}$
c	1	1	$\geq 10^{-6}$–$<3 \times 10^{-6}$
d	2	2	$\geq 10^{-7}$–$<10^{-6}$
e	3	3	$\geq 10^{-8}$–$<10^{-7}$
–	–	4	$\geq 10^{-9}$–$<10^{-8}$

(for continuously operating systems) is for reference only and does not provide for conversion between each standard. For most automated industrial processes, the safety circuit must achieve an ISO 13849 performance level of 'd' or better.

A note of prudence is warranted here, in that the integrity of the safety circuit is one fundamental component of an overall system safety performance. It needs to be commensurate with perspective system as a whole. Identifying a higher-level PL or SIL in a safety circuit does not necessarily achieve a higher level of safety for the system as a whole. Requiring a higher-level PL or SIL for the safety circuit can enter into the realm of diminishing returns, needlessly expending finances, resources, and equipment that could be better invested elsewhere for improved returns on functionality and safety.

As a reference, the objective of the process is to be able to answer YES to all the following questions:

- Have all modes of use (operation, maintenance and service) and reasonably foreseeable events been taken into account?
- Has the method of hierarchy of controls been applied for respective safety measures?
- Have the hazards been eliminated; or are the residual risks from hazards reduced to lowest practicable levels?
- Do the safety control measures implemented NOT create new hazards?
- Are the operator's working conditions NOT compromised or jeopardized by the safety control measures?
- Are the safety control measures complimentary and compatible?
- Are the users sufficiently aware of the inherent hazard(s) to properly use and maintain the safety control measures and are they properly informed of the residual risk(s) for safe use?
- Have reasonably foreseeable events and conditions been accounted for?
- Do the protective measures NOT excessively reduce the ability of the machine to perform its function?

WHEN SHOULD A RISK ASSESSMENT BE DONE?

To be truly effective and provide a constructive means for ensuring workplace safety, a risk assessment should be conducted in the formative timeframe of any undertaking at a facility, such as:

- Before new equipment, processes, or activities are introduced,
- Before changes are introduced to existing processes or activities, including when products, machinery, tools, or equipment change or new information concerning harm becomes available,
- When hazards are identified, or in response to an incident/accident.

Remember, the risk assessment provides the structure and methodology for taking stock and control of potential hazards, which is respective of the task size, value, and/ or scope. To be an effective roadmap for safety, it is best executed in the planning stage of the subject life cycle.

HOW DO YOU PLAN FOR A RISK ASSESSMENT?

In general, determine:

- What the scope of your risk assessment will be, commensurate of the subject scale and scope, such as:
  - If there is a precedent/template in place, ensure it is current.
  - For equipment or process modification, it can be as essential as identification of appropriate standard/regulation which applies
  - For new programs and projects, it can be more comprehensive,
  - Will it produce an auditable means for review over the subject life cycle?
- The resources needed, best accomplished with a review of a training needs assessment (e.g., are team members aware of the hazard(s) and safety requirement(s)).
- What are the metrics and terms? Ensure common core understanding of the principles and definitions.
- Who are the stakeholders involved (e.g., manager, supervisors, engineers, technicians, workers, worker representatives, suppliers, and so on)?
- What relevant laws, regulations, codes, or standards may apply in your jurisdiction, as well as organizational policies and procedures?

HOW IS A RISK ASSESSMENT DONE?

Assessments should be done by a competent person or team of individuals who have a good working knowledge of the situation being studied. Include either on the team or as sources of information, the supervisors and workers who work with the process under review as these individuals are the most familiar with the operation.

In general, to do an assessment, you should:

- Identify hazards.
- Determine the likelihood of harm, such as an injury or illness occurring, and its severity.

- Consider normal operational situations as well as nonstandard events such as maintenance, shutdowns, power outages, emergencies, extreme weather, and so on.
- Review all available health and safety information about the hazard such as Safety Data Sheet (SDS), manufacturers literature, information from reputable organizations, results of testing, workplace inspection reports, records of workplace incidents (accidents), including information about the type and frequency of the occurrence, illnesses, injuries, near misses, and so on.
  - Understand the minimum legislated requirements for your jurisdiction.
- Identify actions necessary to eliminate the hazard, or control the risk using the hierarchy of risk control methods.
- Evaluate to confirm if the hazard has been eliminated or if the risk is appropriately controlled.
- Monitor to make sure the control continues to be effective.
- Keep any documents or records that may be necessary. Documentation may include detailing the process used to assess the risk, outlining any evaluations, or detailing how conclusions were made.

With reference to ANSI B11.0, the process flow, see Figure 11.3.

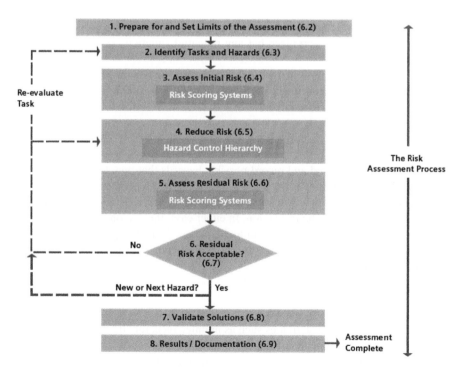

FIGURE 11.3 Risk assessment process flow (ref. ANSI B11.0)

When doing an assessment, also take into account:

- The methods and procedures used in the processing, use, handling, or storage of the substance, and so on.
- The actual and the potential exposure of workers (e.g., how many workers may be exposed, what that exposure is/will be, and how often they will be exposed).
- The measures and procedures necessary to control such exposure by means of engineering controls, work practices, and hygiene practices and facilities.
- The duration and frequency of the task (how long and how often a task is done).
- The location where the task is done.
- The machinery, tools, materials, and so on, that are used in the operation and how they are used (e.g., the physical state of a chemical, or lifting heavy loads for a distance).
- Any possible interactions with other activities in the area and if the task could affect others (e.g., cleaners, visitors, and so on).
- The life cycle of the product, process, or service (e.g., design, construction, uses, decommissioning).
- The education and training the workers have received.
- How a person would react in a particular situation (e.g., what would be the most common reaction by a person if the machine failed or malfunctioned).

It is important to remember that the assessment must take into account not only the current state of the workplace but any potential situations as well (reasonably foreseeable events).

By determining the level of risk associated with the hazard, the employer, and the health and safety committee (where appropriate), can decide whether a control program is required and to what level.

HOW ARE THE HAZARDS IDENTIFIED?

Overall, the goal is to find and record possible hazards that may be present in your workplace. It may help to work as a team and include both people familiar with the work area, as well as people who are not—this way you have both the experienced and fresh eye to conduct the inspection. In either case, the person or team should be competent to carry out the assessment and have good knowledge about the hazard being assessed, any situations that might likely occur, and protective measures appropriate to that hazard or risk.

To be sure that all hazards are found:

- Look at all aspects of the work.
- Include non-routine activities such as maintenance, repair, or cleaning.
- Look at accident/incident/near-miss records.
- Include people who work off-site either at home, on other job sites, drivers, teleworkers, with clients, and so on.

- Look at the way the work is organized or done (include experience of people doing the work, systems being used, and so on).
- Look at foreseeable unusual conditions (for example: possible impact on hazard control procedures that may be unavailable in an emergency situation, power outage, and so on).
- Determine whether a product, machine, or equipment can be intentionally or unintentionally changed (e.g., a safety guard that could be removed).
- Review all of the phases of the life cycle.
- Examine risks to visitors or the public.
- Consider the groups of people that may have a different level of risk such as young or inexperienced workers, persons with disabilities, or new or expectant mothers.

In the absence of any guidance or mature structure available for identifying hazards, a checklist can help. The purpose of a checklist is to account for most nominal steps/ hazards; they cannot be all encompassing. When utilized by a competent person, they can serve to help trigger or identify obscure or secondary hazards requiring attention. One must be mindful not to haphazardly run through "checking off" the items, but to work through them and give due consideration.

Competent person(s) assigned to RA for respective hazard groups can help identify the individual hazards of:

- Laser beam and material interactions
- Electrical supply and distribution
- Mechanical, robotic, material handling, automation
- Chemical, material, or substance (oil mist, metal dust, flammable liquids)
- Compressed gases (process gases—Air, Nitrogen, Oxygen, Argon, Helium shield gases/lasing gases—CO_2, He, N_2, Excimer gases)
- Hydraulic and pneumatic
- Environmental (LGACs, noise, vibration)
- Ergonomics and human factors, trip and slip hazards, ventilation and confined spaces
- Emergency response planning is an invaluable tool that could potentially identify unknown hazards not covered by the above
- Fire and explosion prevention and containment measures

How Do You Know If the Hazard Will Cause Harm (Poses a Risk)?

Each hazard should be studied to determine its level of risk. To research the hazard, you can look at:

- Product information/manufacturer documentation.
- Past experience (knowledge from workers, and so on).
- Legislated requirements and/or applicable standards.
- Industry codes of practice/best practices.
- Health and safety material about the hazard such as safety data sheets (SDSs), research studies, or other manufacturer information.

- Information from reputable organizations.
- Results of testing (atmospheric or air sampling of workplace, biological swabs, and so on).
- The expertise of an occupational health and safety professional.
- Information about previous injuries, illnesses, near misses, incident reports, and so on.
- Observation of the process or task.

Remember to include factors that contribute to the level of risk such as:

- The work environment (layout, condition, and so on).
- The systems of work being used.
- The range of foreseeable conditions.
- The way the source may cause harm (e.g., inhalation, ingestion, and so on).
- How often and how much a person will be exposed.
- The interaction, capability, skill, and experience of workers who do the work.

HOW ARE RISKS RANKED OR PRIORITIZED?

Ranking or prioritizing hazards is one way to help determine which risk is the most serious and thus which to control first. Priority is usually established by taking into account the employee exposure and the potential for incident, injury, or illness. By assigning a priority to the risks, you are creating a ranking or an action list.

There is no one simple or single way to determine the level of risk. Nor will a single technique apply in all situations. The organization has to determine which technique will work best for each situation. Ranking hazards requires the knowledge of the workplace activities, urgency of situations, and most importantly, objective judgment.

For simple or less complex situations, an assessment can literally be a discussion or brainstorming session based on knowledge and experience. In some cases, checklists or a probability matrix can be helpful. For more complex situations, a team of knowledgeable personnel who are familiar with the work is usually necessary.

The form of the risk matrix follows function of the hazard and the environment in which the laser is being used. With regards to laser beam hazards, the risk matrix can potentially be different whether it is one that operates in the Retinal Hazard Region (RHR, 400–1400 nm), Ultra-Violet (UV, 180–400 nm), or Mid to Far InfraRed (MIR, 1400–3000 nm; FIR, 3000 nm–1 mm).

As an example, consider this simple risk matrix for a laser operating in the RHR. The rating system for severity of injury (for most Class 4 industrial lasers*) is almost binary because an intermediate retinal injury is problematic to identify. Although, if the scale of trauma and injury to the retina, eye, and person were included, then the following severity levels can be employed:

---

* Per ANSI Z136, Class 4 lasers merit the DANGER and WARNING signal words for area warning signs. Area warning signs incorporate risk assessment to present the possibility and scale of a hazard causing harm. While similar, do no conflate these terms and concepts.

Consequence (severity):

- Catastrophic: Death or permanently disabling injury or illness (unable to return to work). Includes loss of vision in both eyes, loss of multiple limbs, quadriplegic
- Serious: Severe debilitating injury or illness requiring more than first aid (able to return to work at some point). Includes loss of vision in one eye, loss of a limb.
- Moderate: Significant injury or illness requiring more than first aid (able to return to same job)
- Minor: No injury or slight injury requiring no more than first aid (little or no lost work time).

Likelihood of occurrence (probability):

- Very likely: Near certain to occur
- Likely: May occur
- Unlikely: Not likely to occur
- Remote: So unlikely as to be near zero

The cells in Table 11.3 correspond to a risk level, as described in Table 11.4.

For example, let's work through one element of a laser hazard assessment (foregoing the normal utilization of ANSI Z136 of hazard classification with corresponding control measures). Consider a laser marking application, which requires an operator to load a part into a fixture for marking a barcode and serial number onto a part. This has the following characteristics:

Wavelength = 1090 nm, near infrared, invisible, but still a retinal hazard
Power (average) = 25 Watts

TABLE 11.3

The Relationship Between Probability and Severity, A Two Factor, 4 × 4 Risk Matrix

4 x 4 Risk Matrix		Severity of Harm			
		Catastrophic	Serious	Moderate	Minor
Probability of Occurrence of Harm	Very likely	Extreme	High	High	Medium
	Likely	High	High	Medium	Low
	Unlikely	Medium	Medium	Low	Negligible
	Remote	Low	Low	Negligible	Negligible

TABLE 11.4

Description of Risk Levels

Identifier	Color code	Description
Extreme		Immediately dangerous, stop the process and implement controls
High		Investigate the process and implement controls immediately
Medium		Keep the process going; however, a control plan must be developed and should be implemented as soon as possible
Low		Keep the process going, but monitor regularly. A control plan should also be investigated
Negligible		Keep monitoring the process (do not neglect)

Power (peak) = 6000 Watt
Pulse energy = 1 mJ
Individual pulse width = 4 ns
Pulse repetition rate = 1–100 kHz

The laser itself is identified as a Class 4 (per ANSI Z136) hazard, with the following hazard distances given as:

- the Nominal Ocular Hazard Distance (NOHD) is given as 75 m,
- with a corresponding NSHD (Nominal Skin Hazard Distance) given as 15 m,

That the laser beam emission is at the speed of light, is well in excess of human response capabilities to avert a pending hazard threat.

The assessment team reviewed the situation and agrees that working within the hazard zones to load and unload the parts for marking is likely to:

- Cause a retinal injury due to a specular or diffuse reflection during the marking operations. This outcome is similar to a serious severity rating. At this proximity to the hazard source, it can be argued that a potential exists for both eyes to be injured.
- Likelihood of occurrence is high, given the close proximity of the operator to the unprotected hazard zone of the laser and the volume production nature of the work. This criterion is similar to a very likely probability rating.
- Laser protective glasses are not a reliable protective measure given that:
    - This is a volume operation with continual exposure to the hazard, and human factor considerations preclude the reliability of the operator properly using the PPE consistently with a high degree of confidence.

TABLE 11.5
High Risk Levels

4 x 4 Risk Matrix		Severity of Harm			
		Catastrophic	**Serious**	Moderate	Minor
Probability of Occurrence of Harm	**Very likely**	Extreme	**High**	High	Medium
	Likely	High	High	Medium	Low
	Unlikely	Medium	Medium	Low	Negligible
	Remote	Low	Low	Negligible	Negligible

When compared to the risk matrix chart (Table 11.3), these values, shown in Table 11.5, correspond to a high (or extreme) risk level.

The assessment team decides to implement containment measures to fully constrict the laser beam/radiation paths and fields during marking operations in a controlled reliable manner. The safeguarding measures allows the operator to safely and expediently load and unload the station, it also eliminates potential exposure to plant personnel at work nearby. Training of the operator and affected personnel (e.g., maintenance) regarding the nature of the hazard and for maintaining the integrity of the safety control measures ensures a concerted effort for workplace safety.*

WHAT ARE THE METHODS OF HAZARD CONTROL?

Once you have established the priorities, the organization can decide on ways to control each specific hazard. Hazard control methods are often grouped into the following categories, from most effective to least:

- Elimination (including substitution).
- Engineering controls.
- Administrative controls.
- Personal protective equipment.

Always start with the most effective control measure (for due diligence to be at work) until the risk is as low as reasonably practicable and acceptable. If a higher control measure is no longer practicable, then a lower control measure can be applied (Figure 11.4).

---

* Please note that ANSI Z136 simplifies this process by grouping lasers into hazard classes and then applying the respective safety control measures for that hazard class. The purpose of the above exercise is to identify the embedded risk assessment of Z136 for laser safety so that the principle can be understood and applied for those very large, unique, or high-power Class 4 lasers which warrant a more rigorous examination.

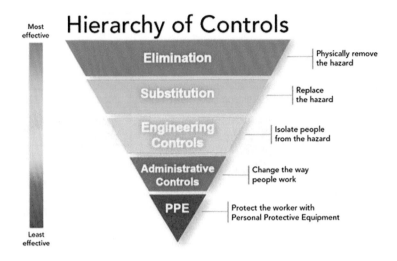

FIGURE 11.4   Hierarchy of controls. (CDC, 2017)

WHY IS IT IMPORTANT TO MONITOR AND REVIEW THE RISK ASSESSMENTS?

Just as a safety circuit must be monitored for its condition status, health, and change of state per cycle, so too must the risk assessment be monitored and reviewed for improvement opportunities.

This falls under the category for the periodic audits (or their equivalent) identified in ANSI Z136 in the normative Appendix A.

The regular review process should serve dual functions, checking the control measures and the risk assessment itself for functionality, currency, and accuracy. Ask if anything has changed in the equipment, or in the knowledge of hazards and exposure limits.

WHAT DOCUMENTATION SHOULD BE DONE FOR A RISK ASSESSMENT?

Keeping records of your assessment and any control actions taken is very important. You may be required to store assessments for a specific number of years. Check for local requirements in your jurisdiction.

The level of documentation or record keeping will depend on:

1. Level of risk involved.
2. Legislated requirements.
3. Requirements of any management systems that may be in place.

Your records should show that you:

- Conducted a good hazard review.
- Determined the risks of those hazards.
- Implemented control measures suitable for the risk.
- Reviewed and monitored all hazards in the workplace.

Suggested Format for a Risk Assessment

Each sector will have its own style for a risk assessment. The following suggestions are provided as a baseline and not intended to preclude any established risk assessment format within an organization or sector (provide example of various matrices).

It is reasonable to expect that the amount of effort required for a risk assessment will be commensurate with the inherent hazard and/or capital value of the project. For those established laser processes, such as marking operations with a certified commercial laser power source, ensuring regulatory conformance to equipment build standards, including US 21CFR1040.10 and 1040.11 or IEC 60825-1 will enable safe use and operation within the workplace. The format for safe use risk assessment of such applications can employ a checklist, structured along the familiar lines of Tables 10 and 11 from ANSI Z136. Example of such is provided below: higher order Class 4 systems (associated with the DANGER signal word).

Industrial applications that employ multi-kilowatt lasers for materials processing are an established technology. In addition to building conformance documentation, and safe use tabulation per ANSI Z136 Tables 10 and 11, a form of system safety data sheet is suggested. Such systems should also include a structured approach for commissioning.

Unique Laser System Applications, Higher Order Class 4

These systems tend to involve purpose built (or customized off-the-shelf) laser power sources for an essentially unique laser materials processing application. Whereas the suggestions for higher order Class 4 risk assessment would provide an appropriate organization and grouping of hazard and risk evaluations, the following approach is recommended:

- Evaluation of the laser power source separate from the application
- Control system safety integrity level and performance requirements identified, with weakest link known. This in turn determines the factor of safety for the design of the laser power source components and its process/application
- Power source evaluation normally part of the design development, to include consideration for:
  - Component build tolerances and potential deviations of construct and their impact on performance and safety.
  - Level of robustness (factor of safety) and component redundancy identified. This includes single fault failure evaluation.
  - Power-up, operational, and shut-down condition monitoring.
  - Environmental deviations, including:
    - Temperature and humidity effects on equipment.
    - Electrical power supply fluctuations, brown-outs, spikes, over and under voltage conditions, power loss (does not necessarily mean that no hazard is possible).
- Process or application evaluation likewise to have same considerations as above for the laser power source, and to include, as applicable:
  - Beam delivery deviations.
  - Second order effects, such as laser-target interactions when thresholds are approached (such as for plasma, ionization, and so on).

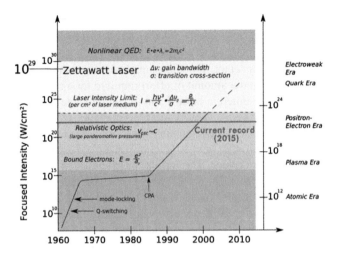

- Process condition monitoring.
- Incremental commissioning with condition monitoring where practicable (not just pass-fail).
- Environmental deviations as noted above.

ATTACHMENT A: EXECUTIVE SUMMARY RISK ASSESSMENT FORM (ONE PAGE EXPLANATION, ONE PAGE FORM)

The following contains an executive level risk assessment form, suitable for those standard systems commonly employed in manufacturing environments.

System description:

LASER POWER SOURCE INFORMATION					
LASER TYPE:		**LASING MEDIUM:**		**WAVELENGTH (nm):**	
MODE: ☐ CW ☐ PULSED	**MAX. AVE. POWER (W):**	**MAX. PEAK POWER (W):**	**PULSE RATE (Hz):**	**PULSE DURATION (s):**	**PULSE ENERGY (J):**
CDRH Accession Number (or Equivalent):					
			US 21CFR1040.10		**ANSI Z136.9**
LASER POWER SOURCE HAZARD CLASS (INHERENT):			IV		4
INTEGRATD SYSTEM LASER HAZARD CLASS (RESIDUAL):			I		1

Residual risk operators and facility personnel ALARP and Acceptable per regulations and facility OHS policy:

OEM Training:

☐ Operation: startup, production, shut-down, E-stops
☐ Maintenance, complete with schedule
☐ Service, complete with contacts
☐ Training roadmap, signed with persons trained and instructors

Authorized Users:

☐ Trained, tested, acknowledged

SOP in place:

☐ Current and approved

LSO Approval

The following attachments are placeholders only. To be upgraded appropriately with watermarks of approval.

ATTACHMENT B: BASIC BUILD SAFETY CONFORMANCE CHECKLIST (PER US 21 CFR 1040.10)

ID:	Mfr:		Customer:				CLASS X
			Following checklist is for the integrated, complete system, which is Class X.				
			Performance Requirements – US FDA/CDRH 21CFR1040.10		Present		
Clause	Class	Description	Comment	Yes	No	N/A	
1040.10(f)(1)	All	Protective Housing	Installation inside unoccupied building. Able to contain, control and endure for intended use and reasonably foreseeable single fault failure conditions	[X]	[ ]	[ ]	
1040.10(f)(2)	All	Safety Interlocks	**Note:** requirements are dependent on class of internal radiation				
		Non-defeatable Interlocks	Access door, gates (per NFPA 79)	[X]	[ ]	[ ]	
		Defeatable Interlocks	No defeatable interlocks used	[ ]	[X]	[ ]	
1040.10(f)(2)(iii)	All	Interlock Failure Protection or Redundancy or Reliability	Dual channel safety circuit with two (2) XXX safety relays, for part presence and shutter. Relays Rated: Cat 3/4 EN954-1, SIL 2/3 IEC 61508, PLd/e ISO 13849-1, SIL 2/3 IEC 62061	[X]	[ ]	[ ]	
1040.10(f)(3)	IIIB, IV	Remote Interlock Connector	With XXX laser power source and system controls.	[X]	[ ]	[ ]	
1040.10(f)(4)	IIIB, IV	Key Control	With XXX system controller	[X]	[ ]	[ ]	
1040.10(f)(5)		Emission Indicator:	Indicators on laser and controls, if separated by more than 2 meters. **Note:** Class IIA is exempt. Class I system does not require emission indicator.				
1040.10(f)(5)	II, III, IV	Emission Indicator	With XXX system controller at station	[X]	[ ]	[ ]	
1040.10(f)(5)(i)	II, IIIA	Emission indicator (no delay)	Not required on Class I system/product	[ ]	[ ]	[X]	
1040.10(f)(5)(ii)	IIIB, IV	Emission Indicator (with delay)	Not required on Class I system/product	[X]	[ ]	[ ]	
1040.10(f)(6)	II, III, IV	Beam Attenuator	With XXX laser per IEC 60825-1 & CDRH LN50	[ ]	[ ]	[X]	
1040.10(f)(7)	II, III, IV	Location of Controls	Per NFPA 79 and industrial best practice	[X]	[ ]	[ ]	
1040.10(f)(8)	All	Viewing Optics	No collecting optics.	[ ]	[X]	[ ]	
1040.10(f)(9)	All	Scanning Safeguard	This is not a scanning laser.	[ ]	[X]	[ ]	
1040.10(f)(10)	IV	Manual Reset Mechanism	Unexpected re-start prevented. Per NFPA 79	[X]	[ ]	[ ]	
No human access into the NHZ is required or possible during operation or maintenance. Service is by authorized OEM XXX technician. Using certified XXX laser product. CDRH A/N: XXXXXXX-XXX							

ID:	Mfr:		Customer:				CLASS X
			Following checklist is for the integrated, complete system, which is Class X.				
			Labelling Requirements – US FDA/CDRH 21CFR1040.10		Present		
Clause	Class	Description	Comment	Yes	No	N/A	
1010.2	All	Certification Label	Per manual, on control cabinet	[X]	[ ]	[ ]	
1010.3	All	Identification Label	Manufacturer tag as required, on cabinet	[X]	[ ]	[ ]	
1040.10(g)(1,2,3)	I Exempt	Class Designation and Warning Label	Per CDRH LN50, Class I identified on safety enclosure/housing. Embedded Class IV laser warning label	[X]	[ ]	[ ]	
1040.10(g)(4)	I, IIA Exempt	Radiation Output Information (Position 2 on label)	On XXX laser power supply In XXX user manuals Referenced in integrator user manual	[X]	[ ]	[ ]	
1040.10(g)(5)	I, IIA Exempt	Aperture Label	Provided on XXX focus optics. Additional with non-interlocked encapsulation housing.	[ ]	[ ]	[X]	
1040.10(g)(6)	All	Non-Interlocked Protective Housing Labels	On exterior housing panels, for service panels. Encapsulation housing/shroud.	[X]	[ ]	[ ]	
1040.10(g)(7)	All	Defeatably Interlocked Protective Housing Labels	No defeatable interlocks are employed.	[ ]	[X]	[ ]	
1040.10(g)(8)	All, as applicable	Warning for Invisible Radiation	Appropriate text where applicable	[X]	[ ]	[ ]	
1040.10(g)(9)	All	Positioning of Labels	Compliant, legible, prominent	[X]	[ ]	[ ]	
1040.10(g)(10)	All	Label Specifications	Per CDRH LN 50, IEC 60825-1	[X]	[ ]	[ ]	
Labelling per CDRH Laser Notice 50 (LN50), IEC 60825-1 format, ANSI Z535 & ISO 3864 compliant. Using certified XXX laser product. CDRH A/N: XXXXXXX-XXX							

ID:		Mfr:	Customer:		CLASS X	
Following checklist is for the integrated, complete system, which is Class X.						

Information Requirements – US FDA/CDRH 21CFR1040.10				Present		
Clause	Class	Description	Comment	Yes	No	N/A
1040.10(h)(1)	All	User Information	Per established industrial practice and integrator quality systems	[X]	[ ]	[ ]
1040.10(h)(1)(i)	All	Operator & Maintenance Instructions	Instructs on safe use and maintenance. Provides maintenance schedule	[X]	[ ]	[ ]
1040.10(h)(1)(ii)	All	Statement of Parameters	Identifies limits for safe use. Includes *"Caution – Use of controls or adjustments or performance of procedures other than those specified herein may result in hazardous radiation exposure"* statement.	[X]	[ ]	[ ]
1040.10(h)(1)(iii)	All	Label Reproductions	Laser safety warning and informational label types, quantities and locations (tagging roadmap) provided in user manual.	[X]	[ ]	[ ]
1040.10(h)(1)(iv)	All	Listing of Controls, Adjustments, and Procedures, including Warnings	User manual explains how the location, operation and adjustment of controls for the laser product is such that human exposure to laser or collateral radiation in excess of the accessible emission limits of Class I (per CDRH, IEC ANSI) is prevented during operation or maintenance work.	[X]	[ ]	[ ]
1040.10(h)(2)(ii)	All	Service Information	Identifies that service work is to be effected only by authorized OEM XXX service technicians	[X]	[ ]	[ ]

Note: Respective manuals and documents from original equipment provider are utilized and referenced. Above information requirements have also been summarized and provided for in the integrator User Manual and SOP for the end customer. See also documentation of user training by integrator for customer, as part of ownership transfer.

Using certified XXX laser product. CDRH A/N: XXXXXXX-XXX

ID:	Mfr:	Customer:	CLASS X
Following checklist is for the integrated, complete system, which is Class X. This identifies accepted substitutions for the CDRH per LN50			

Laser Notice 50 Conformance – IEC 60825-1, IEC 60601-2-22 with 21CFR1040.10			
Requirement	US 21CFR1040 Reference, Applicability	IEC 60825-1:2007 Reference, Applicability	Note
Definitions	(b) All laser products	3 All laser products	
Classification	(c)(1) All laser products	4, 4.3 All laser products	
Accessible emission limits	(d) All laser products	5 All laser products	
Tests for determination of compliance	(e) All laser products	5.1, 5.2, 5.3 All laser products	
Protective housing	(f)(1) All laser products	4.2 All laser products	
Safety interlocks	(f)(2) All laser products	4.3 For human access to Class 3R, Class 3B or Class 4 levels	
Remote Interlock connector	(f)(3) Class IIIb and Class IV laser systems	4.4 Class 3B and Class 4 laser systems	
Key control	(f)(4) Class IIIb and Class IV laser systems	4.6 Class 3B and Class 4 laser systems	
Laser radiation emission indicator	(f)(5) Class II, Class III, and Class IV laser systems	4.7 Class 3B and Class 4 laser systems	
Beam attenuator	(f)(6) Class II, Class III, and Class IV laser systems	4.8 Class 3B and Class 4 laser systems	
Location of controls	(f)(7) Class IIa, Class II, Class III, and Class IV laser products	4.9 Applies to possible exposure to Class 3R, 3B or Class 4 levels	
Viewing optics	(f)(8) All classes of laser products	4.10 All classes of laser products	
Scanning safeguard	(f)(9) All classes of laser products	4.11 All classes of laser products	
Labeling requirements	(g) All laser products	7 All laser products	
User information	(h)(1) All laser products	8.1 All laser products	
Medical laser products	1040.11 (a)	9.2	Not Applicable.

Note: This is an industrial laser marking system, Class I. Embedded with a Class IV certified industrial laser marking power supply.

Using certified XXX laser product. CDRH A/N: XXXXXXX-XXX

ID:	Mfr:	Customer:		CLASS X
Following checklist is for the integrated, complete system, which is Class X. This identifies the applicable sections of 21CFR unaffected by Laser Notice 50 (obligatory compliance)				

Laser Notice 50 Conformance – IEC 60825-1, IEC 60601-2-22 with 21CFR1040.10			
Requirement	**US 21CFR1040 Reference, Applicability**		**Note**
Certification	1010.2	See user manual	Fulfilled
Identification	1010.3	See user manual	Fulfilled
Variances	1010.4	No variances involved	Fulfilled
Applicability	(a)	See user manual	Fulfilled
Removable laser system	(c)(2)	This does **not** incorporate a removable laser system.	Compliant
Manual reset mechanism	(f)(10)	Per NFPA 79	Fulfilled
Purchasing and servicing information	(h)(2)	User replacement information provided in manual. Instructions include information that service is by authorized OEM Osela technicians.	Fulfilled
Modification of a certified laser product	(i)	This does **not** modify a certified laser product.	Compliant
Surveying, leveling and alignment laser products	1040.11 (b)	This is **not** a surveying, leveling or alignment laser product	Compliant
Demonstration laser products	1040.11 (c)	This is **not** a demonstration laser	Compliant
Note: This is an industrial laser marking system, Class I. Embedded with a Class IV certified industrial laser marking power supply. Using certified XXX laser product. CDRH A/N: XXXXXXX-XXX			

ATTACHMENT C: BASIC USER SAFETY CONFORMANCE CHECKLIST (PER ANSI Z136)

ANSI Z136.1 (2014) Table 10a
Control Measures for the Seven Laser Classes

Engineering Control Measures	Classification							
	1	1M	2	2M	3R	3B	4	4U
Protective Housing (4.4.2.1) (Impracticable, other measures required)	X	X	X	X	X	X	X	X
Without Protective Housing (4.4.2.1.1) (Alternative controls established by XXX)	LSO shall establish Alternative Controls							X
Interlocks on Removable Protective Housings (4.4.2.1.3) (N.A. No protective housing.) (Limited open beam path condition, Class 4.)	∇	∇	∇	∇	∇	X	X	∇
Service Access Panel (4.4.2.1.4) (N.A. No protective housing.)	∇	∇	∇	∇	∇	X	X	∇
Key Control (4.4.2.2) (With XXX and XXX units, station has LO-TO capability as well)	—	—	—	—	—	•	•	X
Viewing Windows, Display Screens and Diffuse Display Screens (4.4.2.3) (PPE req'd in LCA) (No portals or displays. See User Manual)	Assure viewing limited < MPE							PPE
Collecting Optics (4.4.2.6) (N/A. No collecting optics employed at facility)	X	X	X	X	X	X	X	N.A.
Fully Open Beam Path (4.4.2.7.1) (NHZ completed. LCA in effect.)	—	—	—	—	—	X NHZ	X NHZ	X
Limited Open Beam Path (4.4.2.7.2) (NHZ completed. LCA in effect.)	—	—	—	—	—	X NHZ	X NHZ	X
Enclosed Beam Path (4.4.2.7.3)	None is required if 4.4.2.1 and 4.4.4.1.3 are fulfilled							N.A.
Remote Interlock Connector (4.4.2.7.4) (With XXX and XXX units for safety interlock and e-stop circuit)	—	—	—	—	—	•	X	X
Beam Stop or Attenuator (4.4.2.7.5) (XXX, XXXX, per IEC 60825-1, CDRH LN 50)	—	—	—	—	—	•	X	X
Area Warning Device (4.4.2.8) (To be established, per Alternative Controls and LSO approval.)	—	—	—	—	—	•	X	—
Class 4 Laser Controlled Area (4.4.2.9 and 4.4.3.5) (N.A. Class 1 machine)	—	—	—	—	—	—	X	—
Protective Barriers and Curtains (4.4.2.5) (Designed for partial containment of NHZ where possible, to restrict human access into LCA.)	—	—	—	—	—	•	•	X

LEGEND	X Shall	MPE Shall if MPE is exceeded
	• Should	NHZ Nominal Hazard Zone analysis required
	— No requirement	* May apply with use of optical aids
	∇ Shall if enclosed Class 3B or Class 4	

ANSI Z136.1 (2014) Table 10b
Table 10b. Control Measures for the Seven Laser Classes (cont.)

Administrative (and Procedural) Control Measures	Classification							
	1	1M	2	2M	3R	3B	4	4U
Standard Operating Procedures (4.4.3.1) (Provided per industrial safety awareness)	—	—	—	—	—	•	X	X
Output Emission Limitations (4.4.3.2) (Assessed, not practicable)	—	—	—	—	—	LSO Determination		X
Education and Training (4.4.3.3) (Provided per industrial safety awareness)	—	•	•	•	•	X	X	X
Authorized Personnel (4.4.3.4) (Identified per XXX H&S policy)	—	—	—	—	—	X	X	X
Indoor Laser Controlled Area (4.4.3.5) (Established as required)	—	*	—	*	—	X NHZ	X NHZ	X
Class 4 Laser Controlled Area (4.4.2.9 and 4.4.3.5) (Applies and established)	—	—	—	—	—	—	X	X
Temporary Laser Controlled Area (4.4.3.10) (Service only by authorized technician and coordinated with customer H&S/LSO)	∇ MPE	∇ MPE	∇ MPE	∇ MPE	∇ MPE	—	—	∇ MPE
Controlled Operation (4.4.3.5.2.1) (Per industrial automation best practices)	—	—	—	—	—	—	•	X
Outdoor Control Measures (4.4.3.6) (N.A. Indoor, industrial Class 4 system)	NHZ	* NHZ	X NHZ	* NHZ	X NHZ	X NHZ	X NHZ	N.A.
Laser in Navigable Airspace (4.4.3.6.2) (N/A. NHZ does not extend into Navigable Airspace, ref. hazard assessment)	•	•	•	•	•	•	•	N.A.
Spectators (4.4.3.7) (N/A. Controlled access facility.)	—	*	—	*	—	•	X	X
Alignment Procedures (**4.4.3.8**) **(per SOP)**	∇	X	X	X	X	X	X	X
Service Personnel (**4.4.3.9**) **(per SOP)**	LSO Determination							∇

Items in red in the right-hand column are to be addressed.

LEGEND	X Shall	MPE Shall if MPE is exceeded
	• Should	NHZ Nominal Hazard Zone analysis required
	— No requirement	* May apply with use of optical aids
	∇ Shall if enclosed Class 3B or Class 4	

The following two summary checklists, from ANSI Z136.1 Table 10c and 10d, identifies support PPE and special considerations for this application. The low power XXX Class 4 laser does not represent a skin hazard (see Hazard Assessment file). Operating conditions are for intended use by authorized and trained personnel for industrial and/or manufacturing environments.

ANSI Z136.1 (2014) Table 10c

Table 10c. Control Measures for the Seven Laser Classes (cont.)

Personal Protective Equipment (PPE)	Classification							
	1	1M	2	2M	3R	3B	4	4U
Laser Eye Protection (4.4.4.2)	—	—	—	—	—	X	X	X
Skin Protection (4.4.4.3)	—	—	—	—	—	•	•	N.R.
Protective Clothing (4.4.4.1, 4.4.4.3 and 4.4.4.3.1)	—	—	—	—	—	•	X	N.R.

LEGEND: X   Shall
              •   Should
              —   No requirement
              N.R. Assessed, not required.

Note: The Class 1 laser marking station fully contains the high power laser marking operations, where laser rated PPE is not required, and special considerations are also not required.

Table 10d. Control Measures for the Seven Laser Classes (cont.)

Control Measures: Special Considerations and Warning Signs	Classification							
	1	1M	2	2M	3R	3B	4	4U
Demonstration with General Public (4.5.1) Not accessible to general public.	—	*	X	*	X	X	X	X
Laser Optical Fiber Transmission Systems (4.5.2) Not applicable.	MPE	MPE	MPE	MPE	MPE	X	X	N.A.
Laser Robotic Automated Installations (4.5.3) (Automated machine, per ANSI/RIA 15.06:2012, as applicable)	—	—	—	—	—	X NHZ	—	X
Laser Controlled Area Warning Signs (4.6)	—	—	—	—	—	X	X	X

LEGEND: X   Shall
              •   Should
              —   No requirement
              *   May apply with use of optical aids
              MPE Shall if MPE is exceeded

ANSI Z136.1 (2014) Table 11a

Note: A Class 1 laser product/machine does not present any laser hazard to personnel, and as such does not require area warning devices and/or signs. The following summary checklist identifies the user safety achieved with the Class 1 laser machine/product for intended use by the authorized and trained operator.

Table 11a. Summary of Area Warning Devices and Signs

Clause	Title	Classification							Required Statement or Comment
		1	2	2M	3R	3B	4	4U	
3.5.1	Personnel	—	√	√	√	√	√	√	Some individuals may be unable to read or understand signs
4.4.2.8.1	Visible Warning Devices	—	—	—	—	√	√	√	Visible warning should be required for Class 3B and shall for Class 4
4.4.2.8.2	Audible Warning Devices	—	—	—	—	√	√	√	Audible warning should be required for Class 3B and shall for Class 4
4.6.1	Design of Signs	—	√	√	√	√	√	√	Per ANSI Z535 requirements
4.6.1.1	Safety Alert Symbol	—	√	√	√	√	√	√	The alert symbol is required on all Caution, Warning and Danger signs
4.6.1.2	Laser Radiation Hazard Safety Symbol	—	√	√	√	√	√	√	The laser sunburst is required on all signs per ANSI Z535
4.6.1.3	Laser Warning Sign Posting	—	—	—	√	√	√	√	Specifies posting requirements of Class 3R, Class 3B and Class 4
4.6.1.4	Laser Warning Sign Purpose	—	—	—	—	√	√	√	States the four purposes of area warning signs
4.6.2	Area Warning Sign Signal Words	—	√	√	√	√	√	√	Specifies which sign is required: Danger, Warning, or Caution
4.6.2.1	Signal Word "Danger"	—	—	—	—	—	√	√	Specifies when to use the word "Danger" and format
4.6.2.2	Signal Word "Warning"	—	—	—	—	√	√	√	Specifies when to use the word "Warning" and format
4.6.2.3	Signal Word "Caution"	—	√	√	√	—	—	—	Specifies when to use the word "Caution" and format
4.6.3	Pertinent Sign Information	—	√	√	√	√	√	√	Specifies the format of signs
4.6.3.4	Message Panel Information	—	√	√	√	√	√	√	Specifies the wording of message panel
4.6.4	Location of Signs	—	√	√	√	√	√	√	Specifies the location of signs

NOTE—Signs and labels prepared in accordance with Z136.1-2007 (and prior editions) are considered to fulfill the requirement of the standard.
LEGEND— √ denotes that the section applies to the applicable Class of laser.

ANSI Z136.1 (2014) Table 11b

Table 11b. Summary of Labeling Requirements

Clause	Title	Classification						Required Statement or Comment
		1	2	3R	3B	4	4U	
3.5.1	Personnel	√	√	√	√	√	√	Some individuals may be unable to read or understand labels
4.4.2.1	Protective Housing	√	√	√	√	√	√	Specific word depending on internal laser Class
4.4.2.1	Conduit Label	—	√	√	√	√	√	
4.4.2.1.4	Service Access Panel	√	√	√	√	√	√	Label required if removal permits access to laser
4.4.2.1.5	Equipment Label Information	√	√	√	√	√	√	Specifies specific wording by Class
4.5.2 4.5.2.1	Optical Fiber Transmission	—	—	√	√	√	√	Words required if the disconnect is not located in a laser controlled area
4.6.6	Warning Label	—	√	√	√	√	√	Class label with symbols and specific words (per IEC and CDRH LN #50)

NOTE 1—Labeling of laser equipment in accordance with the Federal Laser Product Performance Standard (FLPPS) or IEC 60825-1 may be used to satisfy the equipment labeling requirements in this standard.
NOTE 2—Signs and labels prepared in accordance with Z136.1-2007 (and prior editions) are considered to fulfill the requirement of the standard.
NOTE 3—See also Appendix D, Table D2 for additional applicable labeling information.
LEGEND— √ denotes that the section applies to the applicable Class of laser.

Table 11c. Summary of Protective Equipment Labeling

Clause	Title	Summary
4.4.4.2	Protective Eyewear	OD and wavelength marking required
4.4.2.3	Viewing Windows and Display Screens	OD, wavelength and exposure time marking recommended
4.4.2.4	Facility Windows	OD, wavelength and exposure time marking required
4.4.2.5	Protective Barrier	Threshold limit and exposure time marking required, see Appendix C2.4
4.4.2.6	Collecting Optics Filters	OD, wavelength and threshold limit marking required

NOTE 1—Signs and labels prepared in accordance with Z136.1-2007 (and prior editions) are considered to fulfill the requirement of the standard.
NOTE 2—Labeling is only required when windows, filters or barriers are not sold as an integral part of the product.
NOTE 3—See also Appendix D, Table D2 for additional applicable labeling information.

ATTACHMENT D: EXAMPLE LASER SAFETY DATA SHEET

LASER SAFETY & HAZARD EVALUATION FORM		
COMPANY/FACILITY	**LASER SYSTEM LOCATION**	**LASER SAFETY OFFICER (LSO)**
BUILD LASER SAFETY		**USER LASER SAFETY**
NATIONAL ☐ U.S. 21CFR1040.10 (FLPPS) ☐ IEC 60825-1 ☐		**NATIONAL** ☐ ANSI Z136.9 ☐ IEC 60825-14 ☐
STATE		**STATE**
LOCAL		**LOCAL**
FACILITY		**FACILITY**

INTEGRATED SYSTEM INFORMATION							
MANUFACTURER NAME:				**PHONE:**			
ADDRESS:				**FAX:** **WEB:**			
PRODUCT NAME/ID:				**DATE OF MANUFACTURE:**			
MODEL NUMBER:				**SERIAL NUMBER:**			
CONFORMANCE TO FLPPS:		☐ YES		☐ NO		☐ N/A	
ACCESSION No.:				**VARIANCE No.:**			
CLASSIFICATION:	☐ 1	☐ 2	☐ 2A	☐ 3A	☐ 3B	☐ 4	
CONFORMANCE TO IEC 60825-1		☐ YES		☐ NO		☐ N/A	
CLASSIFICATION:	☐ 1	☐ 1M	☐ 2	☐ 2M	☐ 3R	☐ 3B	☐ 4

EMBEDDED LASER INFORMATION							
MANUFACTURER NAME:				**PHONE:**			
ADDRESS:				**FAX:** **WEB:**			
PRODUCT NAME/ID:				**DATE OF MANUFACTURE:**			
MODEL NUMBER:				**SERIAL NUMBER:**			
CONFORMANCE TO FLPPS:		☐ YES		☐ NO		☐ N/A	
ACCESSION No.:				**VARIANCE No.:**			
CLASSIFICATION:	☐ 1	☐ 2	☐ 2A	☐ 3A	☐ 3B	☐ 4	
CONFORMANCE TO IEC 60825-1		☐ YES		☐ NO		☐ N/A	
CLASSIFICATION:	☐ 1	☐ 1M	☐ 2	☐ 2M	☐ 3R	☐ 3B	☐ 4

LASER POWER SOURCE INFORMATION					
LASER TYPE:		**LASING MEDIUM:**		**WAVELENGTH (nm):**	

MODE: ☐ CW ☐ PULSED	**MAX. AVE. POWER (W):**	**MAX. PEAK POWER (W):**	**PULSE RATE (Hz):**	**PULSE DURATION (s):**	**PULSE ENERGY (J):**

BEAM DIAMETER AT EXIT PORT/APERTURE:	**BEAM DIVERGENCE, FULL ANGLE (mrad):**

FEED FIBER SIZE (μm):	**NUMERICAL APERTURE:**	☐ 1/e ☐ $1/e^2$	**BPP (mm*mrad):**
PROCESS FIBER SIZE (μm):	**NUMERICAL APERTURE:**	☐ 1/e ☐ $1/e^2$	**BPP (mm*mrad):**

COLLIMATION LENS, FOCAL LENGTH (mm):	**BEAM DIAMETER ON LENS (mm):**	☐ 1/e ☐ $1/e^2$
OBJECTIVE LENS, FOCAL LENGTH (mm):	**BEAM DIAMETER ON LENS (mm):**	☐ 1/e ☐ $1/e^2$

NOTE: FOR SAFETY CALCULATIONS USE 1/e BEAM DIMENSIONS.
WHERE DIA(1/e) = DIA($1/e^2$)/$\sqrt{2}$
BEAM DIAMETER ON OBJECTIVE LENS:

BEAM PATH - DELIVERY SYSTEM:	☐ OPEN ☐ CLOSED	**BEAM DELIVERY SAFETY INTERLOCKED:**	☐ YES ☐ NO

LASER HAZARD ANALYSIS			

WAVELENGTH (nm)		**EXPOSURE DURATION (s):**	
MPE:	☐ W·cm^{-2} ☐ J·cm^{-2}		
OD:	**NOMINAL HAZARD DISTANCE (m):**	**DIFFUSE HAZARD DISTANCE (m):**	

ACCESSIBLE HAZARD LEVEL:
☐ ANSI Z136.9
☐ IEC 60825-1 (IEC 60825-14)

	1	1M	2	2M	3R	3B	4	COMMENT
OPERATION								Authorized personnel only.
MAINTENANCE								Maintenance by LSO approved personnel + SOPs.
SERVICE								Service by OEM technicians + SOPs

ELECTRICAL HAZARDS: ☐ POWER SOURCE ☐ EXPOSED WIRING ☐ MISSING COVERS ☐ STORED ENERGY ☐	**IS LOCKOUT REQUIRED:** ☐ YES ☐ NO **PROCEDURES ESTABLISHED:** ☐ YES ☐ NO	

LASER HAZARD ANALYSIS – CONTINUED FOLLOWING

LASER HAZARD ANALYSIS - CONTINUED		
LGACs: ☐ CHEMICAL EMISSIONS ☐ METALLIC FUMES ☐ PARTICULATES ☐	**VENTILATION PROVIDED:** ☐ LOCAL EXHAUST ☐ GENERAL VENTILATION ☐ FILTRATION ☐	
FIRE HAZARDS: ☐ IMPROPER BEAM ENCL. ☐ COMBUSTIBLE MATERIALS ☐ GAS/VAPOR IGNITION ELECTRICAL CIRCUITS ☐	**COMPRESSED GAS CYL'RS:** ☐ PROPERLY STORED ☐ RESTRAINED ☐ REQUIRED SIGNAGE ☐ NO MISSING CAPS ☐	**CHEMICAL HAZARDS:** ☐ COOLANTS ☐ SOLVENTS ☐ GASES ☐
OPTICAL HAZARDS: ☐ DISHCARGE TUBES ☐ UV/WELDING ☐ VISIBLE ☐ IR ☐ PLASMA ☐	**ACOUSTIC HAZARDS:** ☐ YES ☐ NO ☐ KNOWN dB(A): ☐	

LASER HAZARD CONTROL MEASURES				
OPERATION:	**AREA SIGNAGE:** ☐ DANGER ☐ WARNING ☐ CAUTION ☐ NOTICE ☐ N/A	**SOPs ESTABLISHED:** ☐ YES ☐ NO ☐ N/A ☐		**COMMENT:** *System designed and build for Class 1 safe use operation by personnel. Laser Safety Awareness training provided to affected personnel.*
MAINTENANCE:	**AREA SIGNAGE (TEMP):** ☐ DANGER ☐ WARNING ☐ CAUTION ☐ NOTICE ☐ N/A	**SOPs ESTABLISHED:** ☐ YES ☐ NO ☐ N/A ☐		**TEMPORARY CONTROLS:** ☐ BARRIERS ☐ CURTAINS ☐ BEAM BLOCKS ☐ PATH COVERS ☐ PPE
SERCVICE:	**AREA SIGNAGE (TEMP):** ☐ DANGER ☐ WARNING ☐ CAUTION ☐ NOTICE ☐ N/A	**SOPs ESTABLISHED:** ☐ YES ☐ NO ☐ N/A ☐		**TEMPORARY CONTROLS:** ☐ BARRIERS ☐ CURTAINS ☐ BEAM BLOCKS ☐ PATH COVERS ☐ PPE

TRAINING PROVIDED: ☐ LSO ☐ OPERATOR ☐ MAINTENANCE ☐	**USER MANUAL:** ☐ INTENDED USE SCOPE ☐ MAINTENANCE SCOPE ☐ SERVICE SCOPE ☐ BEAM ALIGNMENT	**COMMENT:**

LASER PROTECTIVE EYEWEAR (LPE):	☐ AVAILABLE		☐ NOT REQUIRED
LPE MANUFACTURER/SUPPLIER:			
MODEL:	**OD @ λ:**	**VLT %:**	
VIEWING PORTAL:	☐ PRESENT		☐ NOT APPLICABLE
MANUFACTURER/SUPPLIER:			
MODEL:	**OD @ λ:**	**VLT %:**	
REPLACEMENT INFO.:	☐ NOTED IN USER MANUAL	☐ INFO ON PORTAL	

LASER SAFETY CONFORMANCE AUDIT	
USER SAFETY CONFORMANCE AUDIT: ☐ ANSI Z136.9 ☐ IEC 60825-1 (IEC 60825-14)	**AUDIT FORM:**
AUDIT COMPLETED BY:	**AUDIT DATE:**
LSO REVIEW:	**REVIEW DATE:**

☐ **APPROVED**	☐ **DISCONTINUE USE**	☐ **NEW SAFETY SHEET**

NOTE: Any deficiencies are to be documented with corrective actions taken by responsible persons. LSO to review conformance upon completion.

LASER SAFETY AUDIT NOTES AND COMMENTS (IF APPLICABLE):

ATTACHMENT E

Purpose:

This is a summary risk assessment document pertaining to user laser safety, drawn from ANSI Z136.9.

Applicability:

The structure and organization of the ANSI Z136.9 consensus standard for the "Safe Use of Lasers in Manufacturing Environments" is premised upon an assessment of risks for hazard groups with respective safety control measures, identified in Tables 10 and 11, providing a well-defined risk assessment process.

This supports but does not supersede the applicable CDRH or IEC 60825 compliance filing and documentation process. Other non-laser equipment build and user safety standards and regulations will apply.

The following worksheets are:

- One-page system laser hazard class summary (residual and inherent)
- Five-part detailed worksheets, following Z136.9 control measures groups:
    - Engineering (for laser beam related hazards)
    - Administrative and Procedural (for laser beam related hazards)
    - Personal Protective Equipment (for laser beam related hazards)
    - Miscellaneous (for laser beam related hazards)
    - Non-Beam Hazards (for secondary beam hazards and topical equipment considerations)

Color coding is also employed to provide visual guidance on the laser risk/hazard classes.

Laser Hazard Class	4	3B	3R (SLAs)	2	1
Risk/Hazard Potential	High	Moderate	Low	Low	None (ALARA)

These worksheets have been developed for industrial laser systems which utilize a high-power Class 4 laser for materials processing, embedded within a Class 1 system, to enable safe use and operation.

Manufacturer:	System integrator. Design and build.	Residual Hazard Class: 1
LMP Type:	Class 4 laser brazing power supply, embedded into a Class 1 system/enclosure	
Station/Project ID:	Brass/Asset Tag, Operation Number, Station Name	
	Ref. Laser Safety Data Sheet for details	
Laser Power Source:	HyperDiodeLaser GmbH, Germany, HDL 6000-60	Inherent Hazard Class: 4
Model	HDL 6000-60	
S/N	WYSIWYG-314159	
Power, Mode	6600 Watts, Continuous Wave	
Wavelength	980-1060 nm, ± 10 nm, NIR, Invisible	
MPE	3.63×10^{-3} W·cm^{-2}, t=30,000 seconds. Most restrictive. Eye	
OD	7: Calculated as 6.67, simplified method, most restrictive. Eye	
NHZ$_{fiber-break}$	N/A. Safety interlocked	
NHD$_{intrabeam}$	N/A. Safety interlocked focus optic assembly.	
NHD$_{lens-on-laser}$	228 m	
NHD$_{diffuse}$	7.21 m	
Pilot Laser	HDL, coaxial, visible for path programming	Inherent Hazard Class: 2
Power	1 mW, Continuous Wave, Class 2	
Wavelength	650 nm, Visible (red)	
MPE	1.00×10^{-3} W·cm^{-2}, t=30,000 seconds. Most restrictive.	
OD	1: If viewed longer than 0.25 seconds (overcoming aversion response)	
Process Monitoring	HDL, laser process inspection, Class 4	Inherent Hazard Class: 4
Power	6 W, Continuous Wave and Pulsed (modulated)	
Wavelength	808 nm, Invisible	
MPE	1.64×10^{-3} W·cm^{-2}, t=30,000 seconds. Most restrictive.	
OD	4: Calculated as 3.98, simplified method, most restrictive. Eye.	
NHZ$_{fiber-break}$	N/A. Safety interlocked	
NHD$_{intrabeam}$	N/A. Safety interlocked focus optic assembly.	
NHD$_{lens-on-laser}$	72 m	
NHD$_{diffuse}$	0.34 m	

Note on structure of the Risk Assessment Worksheets:

The hazard classification scheme of the ANSI Z136 consensus standard series for the safe use of lasers, provides means to determine the scale of a laser's potential for harm between 4 major groups with their respective control measures to be applied. There are subdivisions within a group according to the hazard potential under certain conditions, for the purposes of industrial materials processing lasers, the subject interest is in taking the most dangerous (Class 4 lasers) with sufficient control measures so that no hazards (Class 1 conditions) are presented to personnel.

- Class 1 lasers are considered incapable of causing a hazard
- Class 2 lasers operate in the visible spectrum, are considered a low risk hazard, with protection afforded by the aversion response
- Class 3R lasers are low risk lasers used in a controlled environment, such as for surveying, levelling or alignment purposes
- Class 3B lasers are considered moderate risk hazard for intrabeam or specular reflection exposure conditions, but not necessarily diffuse exposures
- Class 4 lasers are considered are most hazardous for direct, specular or diffuse exposure conditions, capable of being a fire ignition source. The scale and immediacy of the hazard capability of these lasers are related to their output energy or power.

As a summary guide dealing with laser beam hazards and their control measures, the general form of a risk assessment structure is provided following. It uses the scheme of dealing with an inherent Class 4 laser (associated) hazard, the application of a control measure (standard/regulation) if applicable, to achieve Class 1 condition for the operator/personnel where no further protective measures are required (other than proper training and testing for competency prior to use). Although generalized for industrial laser materials processing, the equivalency can be treated as:

Hazard/Risk Class	Hazard/Risk Level	Visual Code
4	High	
3B	Moderate	
3R (SLAs)	Low	
2	Low	
1	None, ALARP/ALARA	

The reference risk assessment worksheet is premised upon using ANSI Z136.9 for the hazard classification and safety control measures. The hazard class of the inherent laser used for materials processing (such as cutting, welding, cladding) being Class 4 (for any laser ≥ 0.5 Watts of CW power).

For each hazardous element identified, the noted safety control measure can achieve the requisite objective of Class 1 condition: if sufficiently sized for durability and reliability. It may be that one control measure safeguards against multiple hazards. Unless warranted for redundancy requirements, it is not necessary to have multiple layers of protection against a hazard, where one will suffice.

The worksheet does provide means to account for the nominal risk assessment factors of:

- Severity
- Exposure
- Avoidance

These factors are retained to allow for larger organizations to collect a database for those higher value or unique systems that require a more rigorous assessment. But these do not need to be completed for most industrial laser applications, as Z136.9 has already detailed the hazards and their respective control measures to achieve safety. It is a matter of working through each element ascertaining if the noted control measure can be applied, or if it is addressed by another. Documenting that the residual hazard level is Class 1 for each hazardous element with the appropriate control measure will ensure that Class 1 conditions for the system as a whole is achieved.

ANSI Z136.9 identifies for each inherent hazard of a Class 4 laser, whether that safety measure is required (identified as a "shall" in the Standard) or at least be considered (identified as a "should" in the Standard).

The reference worksheet is drawn from Tables 10 in ANSI Z136.9, with the noted clause and a description of the control measure to address the inherent laser beam hazards. The LSO can document the control measure solution applied, where text in italics provides expanded notes on what the control measure is to achieve in principle or what is typically employed in industrial settings. What is presented in the residual hazard columns is for reference only to illustrate what a completed form can be.

Tables 11 in Z136.9 provides for area warning signage and safety labelling requirements. These are minimal for Class 1 laser systems, as they do not pose a hazard to personnel.

Risk Assessment Worksheet

Engineering Control Measures – Part 1/5

Z136.9 Clause	Description	Inherent Hazard/Risk Controls Required	Severity S1	S2	Exposure E1	E2	Avoidance A1	A2	Risk Index (Hazard Class)	Control Measure Solution	Residual Hazard/Risk Severity S1	S2	Exposure E1	E2	Avoidance A1	A2	Risk Index (Hazard Class)	Hazard Class
		4																1
4.4.2.1	Protective Housing	X							4	Engineered safety enclosure for containment of laser beam paths and fields.							1	✓
4.4.2.1.1	Without Protective Housing	X							4	LSO shall establish Alternative Controls as noted in written LSP							1	✓
4.4.2.1.2	Walk-in Protective Housing	X							4	Alt. control measures employed per established industrial practice, compliant to ANSI/RIA 15.06							1	✓
4.4.2.1.3	Interlocks on Removable Protective Housings	X							4	Controls reliable safety interlocks employed on human access doors and interchange dial table							1	✓
4.4.2.1.4	Service Access Panel	X							4	Service access panels are properly labeled and require a tool for removal							1	✓

4.4.2.2	Key Control	•	4	Certified laser employs requisite key control and is incorporated into the system safety interlock circuit. Additional key control not required.	N.R.	✓
4.4.2.3	Viewing Windows, Display Screens and Diffuse Display Screens	X	4	Assure viewing limited < MPE, per documented laser rated portals with required OD	1	✓
4.4.2.5	Protective Barriers and Curtains	•	4	Controlled entry into the Class 4 is such that exposure < MPE and containment barriers are designed for intended use and reasonably foreseeable single fault failure conditions	1	✓
4.4.2.6	Collecting Optics	X	4	No collecting optics are used in this system. Digital monitoring employed.	N.A.	✓
4.4.2.7.1	Fully Open Beam Path	X NHZ	4	Fully contained and controlled beam path.	N.A.	✓
4.4.2.7.2	Limited Open Beam Path	X NHZ	4	Fully contained and controlled beam path.	N.A.	✓
4.4.2.7.3	Enclosed Beam Path	X	4	Further controls not required if 4.4.2.1 and 4.4.2.1.3 are fulfilled	1	✓
4.4.2.7.4	Remote Interlock Connector	X	4	Certified laser employs requisite remote interlock connector and is incorporated into the system safety interlock circuit. Additional RIC not required.	N.R.	✓

					4		N.R.	✔
4.4.2.7.5	Beam Stop or Attenuator		X		4	Not required for the Class 1 system, where human access to Class 4 conditions are prevented	N.R.	✔
4.4.2.8	Area Warning Device		X		4	Not required for the Class 1 system, where human access to Class 4 conditions are prevented	N.R.	✔
4.4.2.9 and 4.4.3.5	Class 4 Laser Controlled Area		X		4	Laser emission warning not required for the Class 1 system, where human access to Class 4 conditions are prevented	N.R.	✔

LEGEND:

X Shall — No requirement

● Should LSP Laser Safety Program

LSO Laser Safety Officer Determination

N.A. Not Applicable, addressed by other mutually exclusive criteria (e.g. 4.4.2.7.1 vs 4.4.2.7.2 vs 4.4.2.7.3)

N.R. Not Required, for the residual hazard class, addressed through integration efforts as noted

NHZ Nominal Hazard Zone analysis required

∇ Shall if enclosed Class 3B or Class 4

✔ See supporting documentation for conformance assessment and information

Risk Assessment Worksheet

A&P Control Measures – Part 2/5

Z136.9 Clause	Description	Inherent Hazard/Risk — Controls Required	Severity S1	Severity S2	Exposure E1	Exposure E2	Avoidance A1	Avoidance A2	Risk Index (Hazard Class)	A&P Measure Solution	Residual — Severity S1	Severity S2	Exposure E1	Exposure E2	Avoidance A1	Avoidance A2	Risk Index (Hazard Class)	Hazard Class
		4																1
1.3.1	Written Laser Safety Program	X							4	A written laser program has been developed and implemented for the safe operation, maintenance and service of the Class 1 system with an embedded Class 4 laser.							1	✓
4.4.3.1	Standard Operating Procedures	X							4	Work instructions in place for operation, with notation on authorized users and maintenance personnel.							1	✓
4.4.3.2	Output Emission Limitations	LSO							4	Limitation on output emission of the laser determined for production requirements and quality objectives. For operator inspection and maintenance, protocols developed for the LSP and SOPs which employ minimum necessary output emission to accomplish the task at hand. Where possible, use of the alignment laser, or appropriate LOTO followed to minimize or avoid human presence in the NHZ.							1	✓

4.4.3.3	Education and Training	X	4	Safety awareness training provided for Operators, safety training for maintenance personnel is commensurate with hazard Class exposure conditions.	1	✓
4.4.3.4	Authorized Personnel	X	4	Authorized personnel for operation, maintenance and service established by the LSO with posting at the system.	1	✓
4.4.3.5	Indoor Laser Controlled Area	X NHZ	4	The Class 1 system enclosure serves as the NHZ during operation. No further area control measures required.	1	✓
4.4.2.9, 4.4.3.5	Class 4 Laser Controlled Area	X	4	The Class 1 system enclosure serves as the NHZ during operation. No further area control measures required.	1	✓
4.4.3.5.2.1	Controlled Operation	•	4	Written Laser Safety Program establishes safe operation, maintenance and service events for the embedded Class 4 laser.	1	✓
4.4.3.6	Outdoor Control Measures	X NHZ	4	Intended use does not allow for outdoor operation, maintenance or service of this system.	1	✓
4.4.3.6.2	Laser in Navigable Airspace	•	4	Intended use does not allow for outdoor operation, maintenance or service of this system.	1	✓

							✓
4.4.3.7	Spectators	X	4	This is a controlled manufacturing environment. General safety awareness training required for ancillary personnel and visitors. Class 1 system does not pose a hazard to personnel.		1	✓
4.4.3.8	Alignment Procedures	X	4	Alignment procedures established for operators and maintenance.		1	✓
4.4.3.9	Service Personnel	LSO	4	Service by OEM (integrator or laser supplier) authorized personnel. Service personnel are to be laser safety trained as identified by the LSP, noted as such in purchase orders for service.		1	✓
4.4.3.10	Temporary Laser Controlled Area	—	4	Temporary LCA may be required for service, determined on an as-needed basis. LSO to review and approve.		1	✓
				Ref. written Laser Safety Program for additional information on above.			

LEGEND:

X Shall

• Should

LSO Laser Safety Officer Determination

— No requirement

LSP Laser Safety Program

N.A. Not Applicable, addressed by other mutually exclusive criteria (e.g. 4.4.2.7.1 vs 4.4.2.7.2 vs 4.4.2.7.3)

N.R. Not Required, for the residual hazard class, addressed through integration efforts as noted

✓ See supporting documentation for conformance assessment and information

NHZ Nominal Hazard Zone analysis required

∇ Shall if enclosed Class 3B or Class 4

Risk Assessment Worksheet

PPE Control Measures – Part 3/5

Z136.9 Clause	Description	Inherent Hazard/Risk — Controls Required	Inherent — Severity S1	S2	Exposure E1	E2	Avoidance A1	A2	Risk Index (Hazard Class)	PPE Measure Solution	Residual — Severity S1	S2	Exposure E1	E2	Avoidance A1	A2	Risk Index (Hazard Class)	Hazard Class
		4																**1**
4.4.4.2	Eye Protection (Laser Protective Eyewear)	X							4	*Class 1 system does not require laser rated eyewear for operators. As determined in the LSP, maintenance and service conditions have appropriate eyewear established.*							1	✓
4.4.4.3	Skin Protection	•							4	*Class 1 system does not require laser skin protection for operators. Maintenance and service conditions have protocols in place which do not place personnel in the diffuse hazard zone; or where intrabeam or specular reflections are reasonably foreseeable. See written LSP.*							1	✓

4.4.4.1, 4.4.4.3, 4.4.4.3.1	Protective Clothing	X			4					1	✔	

Class 1 system does not require laser rated protective clothing. Maintenance and service conditions have protocols in place which do not place personnel in the diffuse hazard zone; or where intrabeam or specular reflections are reasonably foreseeable. See written LSP.

LEGEND:

X Shall	— No requirement
● Should	**LSP** Laser Safety Program

LSO Laser Safety Officer Determination

N.A. Not Applicable, addressed by other mutually exclusive criteria (e.g. 4.4.2.7.1 vs 4.4.2.7.2 vs 4.4.2.7.3)

N.R. Not Required, for the residual hazard class, addressed through integration efforts as noted

✓ See supporting documentation for conformance assessment and information

NHZ Nominal Hazard Zone analysis required

∇ Shall if enclosed Class 3B or Class 4

Risk Assessment Worksheet

Misc. Control Measures – Part 4/5

Z136.9 Clause	Description	Inherent Hazard/Risk — Controls Required	Inherent — Severity S1	S2	Exposure E1	E2	Avoidance A1	A2	Risk Index (Hazard Class)	Misc. Measure Solution	Residual — Severity S1	S2	Exposure E1	E2	Avoidance A1	A2	Risk Index (Hazard Class)	Hazard Class
		4																
4.5.1	Demonstration with General Public	X							4	This is a controlled access manufacturing environment. Class 1 system does not pose a hazard to personnel during operation. See written LSP.							1	1
4.5.2	Laser Optical Fiber Transmission Systems	X							4	High power laser(s) for materials processing are either: - safety interlocked - have a protective housing - be risk assessed for fault exclusion per 7.3 ISO 13848-1:2015							1	✓
4.5.3	Laser Robotic Automated Installations	X NHZ							4	Robotic and automation consensus standards applied per: - ANSI/RIA 15.06 - ANSI B11.19, B11.20, B11.21, B11.25							1	✓

4.6	Laser Controlled Area Warning Signs	X	4	*Appropriate area warning signs established per ANSI Z535 for the Class 1 system as identified in Z136.9, Tables 11. Posing no hazard, area warning signage requirements are minimal.*			1	✓

LEGEND:

X Shall — No requirement　　　　**NHZ** Nominal Hazard Zone analysis required

● Should　　**LSP** Laser Safety Program　　∇ Shall if enclosed Class 3B or Class 4

LSO Laser Safety Officer Determination

N.A. Not Applicable, addressed by other mutually exclusive criteria (e.g. 4.4.2.7.1 vs 4.4.2.7.2 vs 4.4.2.7.3)

N.R. Not Required, for the residual hazard class, addressed through integration efforts as noted

✓ See supporting documentation for conformance assessment and information

Consideration for intended use operational life cycle, and reasonably foreseeable fault conditions will determine appropriate design and build of the laser system for robustness, reliability and durability. Cumulative effects and/or limit states encountered during equipment operation can enable or generate potential second order effects and hazards. The list of non-beam hazards below are not exhaustive, but serve the LSO to ensure appropriate disciplines, consensus standards and regulations are consulted.

As a summary guide dealing with Non-Beam Hazards (NBHs), the risk assessment structure is further **simplified**, but still retains the scheme of dealing with an inherent Class 4 laser (associated) hazard, the application of a control measure (standard/regulation) if applicable, to achieve Class 1 condition for the operator/personnel where no further protective measures are required (other than proper training and testing for competency prior to use).

The NBHs are grouped first dealing with second order optical radiation hazards, following with nominal equipment hazards.

Risk Assessment Worksheet

NBH Control Measures – Part 5/5		Inherent Hazard/Risk				Residual Hazard/Risk			
Z136.9 Clause	**Description**	Controls Required	YES/NO/NA	Risk Index (Hazard Class)	**NBH Measure Solution**	Audit/Inspected	Pass/Fail/NA	Risk Index (Hazard Class)	Hazard Class
	These non-beam hazards are topical and for reference/guidance only. More comprehensive and applicable consensus standards and regulations will apply.				Identify, if applicable, the appropriate control measure(s) taken to address the noted hazard.				
		4		4					1
7.2.2	Non-Laser Radiation (NLR): *Measures taken to address collateral, optical process, EM, RF, plasma, and/or ionizing radiation associated with LTIR or from the resonator operation.*	App			*Identify those NLR hazards generated/present with appropriate control measures taken.* Ref.: *ACGIH, NIOSH, ICNIRP, 29CFR1926.54, ANSI Z136.9-2013 (G4.2)*			1	1
7.2.2.1.2	LTIR (UV & Blue-Light): *Macro laser materials processing interactions can generate optical re-radiation at the laser-target interaction zone. Potential for significant emissions in the 180-550 nm spectrum. High levels of UV can also produce significant amounts of ozone.*	App			Ref.: - *NIOSH 78–138* - *ANSI/IESNA RP-27* - *ANSI Z136.1-2007 (4.6.2.5.3)* - *ANSI Z136.9-2013 (G4.2.2)* - *ANSI Z87.1, IEC 62471*				✓

7.2.1.2	**LTIR Optical/Process (VIS):** *Macro laser materials processing interactions can generate optical re-radiation at the laser-target interaction zone. Potential for significant emissions in the visible (400-700 nm) spectrum, requiring appropriate attenuation measures for direct viewing.*	App	*Ref.:* - *ANSI/IESNA RP-27* - *IEC 62471* - *ANIS Z136.9-2013 (G.4.2.2)*
	LTIR Thermal (NIR): *The thermal re-radiation of the laser-target interaction can present durability issues on nearby tooling and equipment, presenting a potential for failure.*	App	*E.g. laser cell tooling designed for and equipment specified to meet MIG welding best practices.*
	LTIR Plasma: *High intensity interaction of the laser beam for materials processing can liberate material in the form of a vapor plume, in which further sustained interaction can elevate the plume temperature to a higher level generating plasma.*	App	*E.g. control measures to address the effects of plasma radiation to safeguard personnel.* *Ref. Z136.9-2013 (G1.2), (G4.2.1, G4.2.2)*
	LTIR Ionizing (incl. Bremsstrahlung): *This is the next order of LTIR beyond plasma effects from higher intensity beams on the order of 10^{16} W·cm^{-2}.*	App	*E.g. Principles, objectives and requirements for ionizing radiation control measures found in 21CFR1020.40, 29CFR1910.1096 Health effects: OSHA Directive TED 01-00-015 [TED 1-0.15A], (January 20, 1999). Ref. ANSI Z136.9-2013 (G4.2.1),*

	Item	Type		Score	Reference			✓
	Resonator, Collateral: *The (collateral) pump energy necessary for and/or the byproduct of resonator operations can generate incoherent optical radiation requiring appropriate control measures. These are grouped into the following domains:* - *Optical (e.g. pump diodes and arc/flash lamps)* - *Radio Frequency & Micro Wave (e.g. for excitation)* - *Power Frequency (Extremely Low Frequency) (e.g. with electrical excitation)* - *Plasma (such as with lasing gases)* - *Ionizing (High voltage (> 15 kV) power supplies can generate ionizing radiation)*	App			*E.g. nominally addressed through containment measures in the construct of compliant/certified laser (resonator). User manual should have appropriate inherent hazard information, respective control measures for safety awareness and maintenance, with identification of any residual hazard and additional PPE (or other control measures) if necessary for the operator.* - *Optical Ref.: ANSI/IESNA RP-27, IEC 62471* - *RF, ELF Ref.: IEEE C95.1, C95.6, C95.7* - *Plasma Ref.: ANSI Z136.9-2013 (G1.2, G4.2.1, G4.2.2)* *Ionizing Ref.: ANSI Z136.9-2013 (G4.2.1), OSHA Directive TED 01-00-015 [TED 1-0.15A], (January 20, 1999)*			
7.2.1 7.2.1.4	**Electrical:** *Electrical supply, distribution and control system to applicable safety standard.*	X	YES	4	*Ref.:* - *29 CFR 1910 Subpart S, 29 CFR 1910.147* - *NFPA 70 & 79, UL 508* - *IEC 60204-1* - *ANSI Z136.9-2013 (G4.1)*		1	✓
	Controls safety architecture: *Controls safety integrity/performance level commensurate with the hazard and work environment.*	App		4	*Ref.:* - *NFPA 79, UL 698* - *ANSI B11.TR6* - *EN954-1 (withdrawn)* - *IEC 61508, ISO 13849-1, IEC 62061*		1	✓

					✓
					1

#	Item	Type		Detail
7.2.5	Misc. electrical: Ensuring electrical build conformance will enable safe use. Reasonably foreseeable fault conditions may include: - Direct contact with live circuits, such as during maintenance and service - Indirect contact with circuits which have become active under faulty conditions - Approach to live circuits under high voltage (arc-flash) - Electrostatic phenomena - Resistive heating of circuit components under failure conditions	App	4	During design review(s), ensure that such topics are considered, assessed and addressed. Ref.: - ANSI B11.TR3, B11.0
	Mechanical, Associated with Robots and Automated Equipment: Automated equipment, including robots are to be rendered safe when in the presence of humans or have appropriate safeguards.	App	4	Normally addressed with level 'B' and 'C' standards. Ref.: - ANSI/RIA 15.06 - ANSI B11.19, B11.20, B11.21, B11.25
	General & misc. mechanical: Ensuring electrical build conformance will enable safe use. Reasonably foreseeable fault conditions may include a variety of misfeeds or loss of sequencing in operations which can lead to a loss of hazard(s) containment or pose new hazards to personnel.			Principles include containment, control and guarding measures as identified in: - 29CFR1910.212 (General guarding requirements for all machines) - ANSI B11.0, B11.TR3
7.2.3	Fire: Class 4 laser beams represent a fire hazard potential. Identify appropriate measures taken to protect against reasonably foreseeable operational or fault conditions in which a fire can be started.	App	4	E.g. laser cell tooling designed for and equipment specified to meet MIG welding best practices. Ref.: - NFPA 115, NFPA 51B - ANSI Z136.9-2013 (G1.1.2, G4.3)

7.2.4	**Explosion:** Potential sources for explosion are from stored energy (e.g. capacitor banks) or compressed gases (arc lamps, oxygen tanks/lines), particulate accumulation in fume extraction systems, LTIR failures.	App	4	*Ref.:* - *ANSI Z136.9-2013 (G1.1.3, G4.3)*
7.2.6	**Noise:** Laser resonator, laser-target interaction (LTI) can generate excessive noise. In many cases, sound levels will not result in over-exposure, but may be a nuisance that should be addressed.	App	4	*Ref.:* - *29CFR1910.95* - *ANSI Z136.9-2013 (G1.3)*
7.3.1	**Laser Generated Airborne Contaminants (LGACs):** Laser-target interaction may release various gases, fumes and/or particulates. Appropriate industrial hygiene safety control measures are to be applied.	App	4	*Ref.:* - *ANSI Z136.9-2013 (G2.1, G4.4.1)* - *AWS F1.6, F3.2* - *ACGIH Industrial Ventilation Manual*
7.3.2	**Compressed Gases:** Lasing process, assist or shield gases in use for an application will require appropriate safety control measures for their delivery, use and evacuation.	App	4	*Ref.:* - *29CFR1910.101, 29CFR1910.169* - *CGA P 1* - *ANSI Z136.9-2013 (G2.2, G4.4.2)*
7.3.3	**Laser Dyes and Solvents:** The organic dyes and solvents used in dye lasers are hazardous and may be toxic, carcinogenic and/or flammable.	App	4	*Refer to respective SDS.* *Ref.:* - *ANSI Z136.9-2013 (G2.3, G4.4.3)* - *NFPA (30, 45)*

7.3.4	**Assist Gases:** Safe delivery, deployment and evacuation of assist gases requires appropriate safeguarding. The assist gas can also influence the composition of LGACs as well as spectral response of the laser-target process radiation and NLR.	App	4	Ref.: - ANSI Z136.9-2013 (G2.2, G4.4.2)
7.3.5, 7.3.5.1, 7.3.5.2	**Laser Process Area Ventilation, Prioritization:** Local containment and control of LGACs is most effective when practicable.	App	4	Where utilized, exhaust ventilation recirculation is to comply with ACGIH Industrial Ventilation and Fundamentals Governing the Design and Operation of Local Exhaust Systems, ANSI Z9.2.
7.3.6	**Process Isolation:** A combination of process isolation separate from human access barriers may be required to provide a sufficient margin of safety.	App	4	Identify, if applicable, the levels or layers of process isolation required to achieve the necessary margin of safety and compliance.
7.3.7	**Sensors and Alarms:** Active monitoring of system critical safety elements may be required, commensurate with the hazard-risk, such as for hazardous gas cabinets, toxic or corrosive agents and gases, LGAC filters, etc.	App	4	Identification of those mission critical elements that are safety monitored, in a similar manner that safety interlocks are dual channel and monitored.

7.4	**Biological Agents:** For those lasers dealing with biological agents, etc., mapping of all process pathway and transport vectors with appropriate safeguards is required.	App	4	Additional regulations and standards will apply if dealing with biological substances/agents in manufacturing environments. Such will include but are not limited to: ANSI Z136.3 Ref.: - ANSI Z136.9-2013 (G3, G4.5)
7.5.1, 7.5.2, 7.5.3	Ergonomics: Applicable human-machine interaction and interface is to be considered for safe operations. Noted human factor considerations include: - Limited Work Space - Work patterns - Operator postures - Consideration of hand-arm or foot-leg anthropometry - Sufficient local lighting for tasks - Mental overload and underload, stress - Human error, human behavior - Inadequate design, location or identification of manual controls - Inadequate design or location of visual display units	App	4	Ergonomic considerations improve the effectiveness that personnel can interact with equipment and minimize those contributing factors that can lead to an error in judgement regarding safety with the potential for an accident. Ref.: ANSI B11.TR1

| 7.5.4.1 | Laser Disposal:
Considerations for transfer/disposal of the laser system (a hazard source) have personnel and environmental safety obligations as well as possible restricted technology export constraints. | App | 4 | If the laser is being transferred, resold, or donated, it will be considered as "entering into commerce" and is subject to the requirements of all applicable product safety standards and regulations including 21CFR1040.10 and applicable export controls.

Ref.:
- ANSI Z136.9-2013 (7.5.4.1)
- ANSI Z136.8-2012 (4.4.5, 4.7)
- For product compliance:
- 21CFR1040.10 and 1040.11
- 21CFR1010.2 and 1010.3
- 21CFR1002.10
- For export controls and license:
- The Department of Commerce, Export Administration Regulations (EAR – 15CFR 730-774),
- The Department of State, International Traffic In Arms Regulations (ITAR – 22 CFR 120-130)
- The Treasury Department, Office of Foreign Assets Control (OFAC –31 CFR 500599) |

| 7.5.4.2 | **Laser Waste Disposal:**
Ensure conformance with regulatory (local, state, federal) requirements for hazardous laser waste (e.g. LGAC filters). | App | 4 | *The user manual for the laser system should have information regarding its safety life cycle for the owner/operator. The laser would be considered a hazardous waste (electronic plus associated lasing medium and operating fluids).*

Disposal of the laser and its by-products (e.g. LGAC filters, coolant fluids) are subject to applicable local, state and federal regulations.
If disposing of the laser, it may be necessary to dismantle it to allow for appropriate recycling.

Ref.:
- *29CFR1910 Subpart H*
- *29CFR1910 Subpart Z*
- *EPA National Strategy for Electronic Stewardship (NSES)* | |

LEGEND:

X Shall   — No requirement   **NHZ** Nominal Hazard Zone analysis required

● Should   **LSP** Laser Safety Program   ▽ Shall if enclosed Class 3B or Class 4

LSO Laser Safety Officer Determination

App Application dependent, LSO to determine if hazard potential exists

N.A. Not Applicable, addressed by other mutually exclusive criteria (e.g. 4.4.2.7.1 vs 4.4.2.7.2 vs 4.4.2.7.3)

N.R. Not Required, for the residual hazard class, addressed through integration efforts as noted

✓ See supporting documentation for conformance assessment and information

ONLINE RESOURCES

ASSE, *Top 10 Risks for Risk Assessments*, http://www.asse.org/standards/risk-management/top-10-tips-for-risk-assessments.

CCOHS, *OSH Answer Fact Sheets*, https://www.ccohs.ca/oshanswers/hsprograms/risk_assessment.html.

OSH Wiki, *Occupational Safety and Health Risk Assessment Methodologies*, https://oshwiki.eu/wiki/Occupational_safety_and_health_risk_assessment_methodologies.

Praxiom, *OSHAS 18001: Plain English Definitions*, http://praxiom.com/ohsas-18001-definitions.htm.

CONSENSUS STANDARDS AND REGULATION RESOURCES

The following consensus standards and regulations are drawn from to provide the following terms regarding risk assessment for lasers, laser systems, and laser applications:

- ANSI B11.0
- ISO 31000, Risk Management Principles and Guidelines
- ISO Guide 73, Risk Management—Vocabulary
- The Canadian Standards Association (CSA) Z1002 Standard "Occupational health and safety—Hazard identification and elimination and risk assessment and control."

12 Oversight Regulations and Reported Accidents

Ken Barat

CONTENTS

MANUFACTURER OVERSIGHT

The few regulatory agencies that have some laser oversight, either users or manufacturers, are also interested in accidents. In the United States, we are talking about CDRH. They are not so much interested in individual incidents, but in seeing if a trend is being established: Has the same model medical laser been reported in "X" number of incidents or failures? CDRH is the only way for national action to take place. Yes, the manufacturer could recognize a problem and attempt to fix it. But now we are counting on the professionalism of the manufacturer. If product history has taught us anything, it is this is an unreliable metric. Cost factors and management culture are too much of a variable to count on a firm doing the right thing. Of course, this is strictly this author's opinion, but I will bet many feel the same way. There are just too many examples out in the public domain that support my view.

On the odd chance one is not familiar with CDRH, it means the Center for Devices and Radiological Health and is the branch of the U.S. Food and Drug Administration (FDA) responsible for the premarket approval of all medical devices, as well as overseeing the manufacturing, performance, and safety of these devices. But for lasers users, this includes setting product safety rules for electronic products, including lasers and how laser light shows are conducted.

In addition to reviewing product report submissions, the CDRH also has an Office of Compliance (OC). Using the OC's own words, "OC takes targeted, risk-based compliance actions that address significant violations of device-related laws. We also promote public health by facilitating innovation and fostering a culture of quality within an ever-expanding global medical device market."

CDRH, while covering all electronic products, has its strongest emphasis on medical products. Which is why for nonmedical laser products, getting feedback is slow.

On the medical side, there are several groups:

Division of Analysis and Program Operations (DAPO) analyzes data, develops policy, drafts process, collaborates with FDA's Office of Regulatory Affairs (ORA) on inspection planning and assignments, runs the establishment registration and listing program, and supports recall processing and establishment inspection reviews.

Division of Bioresearch Monitoring (DBM) provides regulatory oversight of medical device clinical investigations, nonclinical good laboratory practice, and institutional review boards in support of the premarket review program. The division coordinates and reviews monitoring inspections of regulated parties and takes necessary action when appropriate. It also investigates and coordinates allegations of research misconduct.

Division of International Compliance Operations (DICO) focuses on foreign device manufacturer and importer assessment; international audit program, compliance policy, and guidance development; export operations and policy; and stakeholder communication and outreach.

Division of Manufacturing and Quality (DMQ) leads domestic enforcement activities and recalls related to device quality and safety, and reviews premarket approval application manufacturing sections, site change supplements, and signals and complaints related to product quality. This division is also the lead on device quality policy.

Division of Premarket and Labeling Compliance (DPLC) enforces premarket clearance and approval requirements, as well as labeling and promotion and advertising requirements for medical devices. This division engages in surveillance of industry practices and responds to urgent or high-priority public health concerns, such as fraudulent devices marketed during a pandemic.

The CDRH database MAUDE, contains information on laser accident examples have been already given in this textbook, but here are three typical reports:

EXAMPLE 1: ENDO LASER MODEL: EYELITE PROBLEM: UNINTENDED LASER OUTPUT

The customer reported that during a surgery procedure with an endo laser in repeat mode, the laser fired twice and stopped. When they released the footswitch, the laser fired a third time. Depressing the footswitch again, the same thing happened. They proceeded for 30–40 shots, then the procedure was aborted. They switched to another unit. Patient outcome was not provided. Additional information has been requested.

EXAMPLE 2: TATTOO INCIDENT MEDLITE C6

A tattoo remover was treating a tattoo and switched the laser wavelength from 1064 nm to 650 nm by attaching the 650 nm dye conversion handpiece to the end of the articulated arm. She had the patient change the laser safety eyewear to the correct 650 nm glasses, but forgot to change her eyewear to the correct wavelength. When she stepped on the foot switch, she noticed that the 650 nm light was very bright and immediately stopped the laser by releasing the foot switch. She indicates that she saw about 5 flashes before stopping the treatment. After this, she experienced a headache but had no problems with her vision. She was advised to see a physician to examine her eyes for evidence of damage to the eye. She indicates that the ophthalmologist found no damage and her headache went away the following day.

Feedback from Manufacturer: Reporter acknowledges that she forgot to change her eyewear to the correct wavelength. She indicates that she was trained on the use of the laser and that the instruction included the necessity to change the eyewear when using the polymer dye conversion handpieces. The manual clearly states in several places to wear the correct eyewear when installing the dye conversion handpieces. The adverse event was not caused by a failure of the system or from inadequate training or instructions for use and hence no corrective action is being taken or is necessary.

EXAMPLE 3: MEDLITE C6

In 2006, according to the allegedly injured party, a laser technician practitioner was performing a procedure using a Medlite C6 laser at 6 j/cm2, 1064 nm, 10 hz with a 4 mm spot size while compressing the skin with a glass window to force the blood away from the treatment site to minimize the formation of purpura. She reports that during the procedure, she noticed bright spots that caused her to blink. After the procedure, she noticed that there continued to be a "black" spot or shadow in her central vision similar to what happens after you look directly into a light bulb or the sun and then look away. She further reports that 3 days later, she went to see an ophthalmologist who examined her and found bleeding and referred her to a retinal specialist. Sixteen days later, she sent an email to an independent contractor working with Hoya Con Bio, informing her of the adverse event. The independent contractor reviewed the email for the first time and reported the alleged injury to Hoya for the first time 5 days later.

Manufacturer response: Device not evaluated as there is no defect or problem with the device. The event may be related to a lack of eyewear or the use of incorrect eyewear. The eyewear provided with the C6 laser system is to protect for use with the 1064 and 532 nm wavelengths. This eyewear has a slight amber tint. The reporter requested that a Hoya representative send her a clear set of eyewear without the amber tint. Hoya sent clear eyewear to her, but sent eyewear rated only for 2940 nm. Upon receipt of the eyewear, she did not check the rating on the eyewear prior to using them. She used them for several months prior to November 2006. She claims to have been wearing this eyewear in 2006 when she allegedly received a back reflection off of the surface of the compression window she was using. The 2940 nm eyewear was removed from the treatment room immediately after the event and replacement eyewear with the correct protection was sent to the institute in December 2006.

Example 4: Palm Burn Catalog #840-846

During operation of the holmium laser during a surgical case, the single use 550 duo tome fiber burned through its fiber and insulation causing a burn to the surgeon's palm. The event occurred approx. 20 mins after the laser/fiber delivery system were in use. There was no patient harm.

USER OVERSIGHT

Here we have two levels: Federal and State Agencies.

State Oversight

The states that have laser regulations require users to report laser accidents within 24 hours or sooner. Of course, while this sounds like they are very interested in public safety, the reality is the position is bogus.

As an LSO, one knows it takes a day to gather all the information the state wants to know, so rushing to inform them can work against you. One reason is they might be at your front door within 24 hrs. That just distracts from putting together a coherent investigation. As well as putting extra pressure on all involved.

At the time of publication, here are the most active state programs: Arizona, Texas, Massachusetts, Florida, Illinois, and New York. Most of these states have the Agency responsible for laser oversight within the Department of Health; in New York, it is within the Department of Labor.

As an example, here is the incident reporting requirements from the state of Texas. They require notification within 24 hours and a written report within 30 days.

(z) Notification of Injury Other Than a Medical Event

1. Each registrant of Class 3b or 4 lasers or user of an IPL device shall immediately seek appropriate medical attention for the individual and notify the agency by telephone of any injury involving a laser possessed by the registrant or an IPL device, other than intentional exposure of patients for medical purposes, that has or may have caused:
   a. An injury to an individual that involves the partial or total loss of sight in either eye; or
   b. An injury to an individual that involves intentional perforation of the skin or other serious injury exclusive of eye injury.
2. Each registrant of Class 3b or 4 lasers or user of an IPL device shall, within 24 hours of discovery of an injury, report to the agency each injury involving any laser possessed by the registrant or IPL device possessed by a user, as applicable, other than intentional exposure of patients for medical purposes, that may have caused, or threatens to cause, an exposure to an individual with second or third-degree burns to the skin or potential injury and partial loss of sight.
   a. Reports of injuries.
      1. Each registrant of Class 3b or 4 lasers or user of an IPL device shall make a report in writing, or by electronic transmittal, within 30 days to the agency of any injury required to be reported in accordance with subsection (z) of this section.
      2. Each report shall describe the following:
         a. The extent of injury to individuals from radiation from lasers or IPL devices;
         b. Power output of laser or IPL device involved;
         c. The cause of the injury; and
         d. Corrective steps taken or planned to be taken to prevent a recurrence.
3. Any report filed with the agency in accordance with this subsection shall include the full name of each individual injured and a description of the injury. The report shall be prepared so that this information is stated in a separate part of the report.
4. When a registrant or user of an IPL device is required in accordance with paragraphs (1)-(3) of this subsection to report to the agency any injury of an individual from radiation from lasers or IPL devices, the registrant or user of an IPL device shall also notify the individual. Such notice shall be transmitted to the individual at a time not later than the transmittal to the agency.
   b. Medical event.
      1. The registrant of Class 3b or 4 lasers or user of an IPL device shall notify the agency, by telephone or electronic transmittal, within 24 hours of discovery of a medical event or of any injury to or death of a patient. Within 30 days after a 24-hour notification is made,

the registrant of Class 3b or 4 lasers or user of an IPL device shall submit a written report to the agency of the event.

2. The written report shall include the following:
    a. The registrant's or user's name;
    b. A brief description of the event;
    c. The effect on the patient;
    d. The action taken to prevent recurrence; and
    e. Whether the registrant or user informed the patient or the patient's responsible relative or guardian. (3) When a medical event occurs, the registrant or user shall promptly investigate its cause, make a record for agency review, and retain the records as stated in subsection (ee) of this section.

FEDERAL USER OVERSIGHT

Federal oversight occurs within OSHA, which has a well-established incident reporting mechanism. Many laser incidents get filed under skin or eye injury and the laser competent seems to get hidden. Searching for laser incidents yields a small number. Many are non-beam injuries involving laser equipment, as opposed to laser eye injuries.

Here are some examples from the OSHA website:*

EXAMPLE 1: ACCIDENT: 200841880—EMPLOYEE DIES IN LASER CUTTER ACCIDENT

On June 7, 2010, Employee #1 was operating an automated laser cutter within the production area. The front guard was raised and Employee #1 was making adjustments inside the point or operation area. Employee #1 became caught in a pinch point when the structure that holds the laser cutter moved to the right front area near the metal structure of the machine. Employee #1's head became caught in the limited space between the metal structures, approximately three inches wide. Another coworker heard a noise and found Employee #1 in the machine. The injury was immediately fatal.

EXAMPLE 2: ACCIDENT: 201503661—DAMAGED MACULA

On March 14, 2003, an employee of UC Berkeley, was working with a Class 4 Nd:YAG near an infrared open beam laser. He wore laser safety eyewear during beam alignment, but afterward, he removed safety eyewear. While taking a meter measurement for output energy, his eye caught a flash from a hidden optic in the beam's path. His exposure exceeded current limits for lasers. The exposure to the laser damaged the macula in his retina, resulting in irreversible retinal damage and loss in visual acuity.

---

* OSHA, *Accident Search Results*, https://www.osha.gov/pls/imis/AccidentSearch.search?acc_
keyword=%22Laser%22&keyword_list=on.

EXAMPLE 3: ACCIDENT: 741579—ELECTRIC SHOCK:
DIRECT CONTACT WITH ENERGIZED PARTS

A mechanic who was in charge of servicing laser cutting systems was adjusting the intensity of a laser. The laser was part of an acme/cleveland laser cutting system (model no. 1510-s). the employee, who had received on-the-job training, gained access to the laser's lens by removing an electrical panel on the side of the laser. With the laser energized at 20 kilovolts d-c, the mechanic reached inside and adjusted the lens on the laser with an insulating tool. He was kneeling on the concrete floor next to the laser and was not wearing rubber insulating gloves. As he was doing this, the back of his left hand contacted the end of the laser, and he received an electric shock. The employee was hospitalized for his injury.

EXAMPLE 4: ACCIDENT: 201112380—ELECTRIC SHOCK:
DIRECT CONTACT WITH ENERGIZED PARTS

An employee had discovered an oil leak in a 2000-watt Mazak Turbo laser cutting machine. He informed the shop foreman about the leak and was told to shut the machine down until it could be repaired. He was also told by the vice president of the company to determine where the leak originated. Over the next week, he looked for the source of the leak.

He climbed up on a platform on a side of the machine to look inside. He removed the laser cabinet panel covers and disabled the interlocks so that the machine could be running while he checked for the leak. He found that some hoses were leaking. The vice president told him to replace the hoses. After replacing the hoses, the employee discovered that the machine was still leaking oil. With the machine running, he climbed back up on the platform to determine the source of the leak. He discovered that the laser resonator unit had oil on it. He was told to shut down the machine until a part could be ordered. He shut down the machine, but did not replace the laser cabinet panel covers or enable the interlocks.

A week after the employee discovered the leak, the part (an O-ring for the laser resonator unit) had arrived and another person came to replace it. The employee showed him where the machine was leaking oil and then went to lunch. When the employee returned from lunch, the vice president told him that the O-ring had been replaced and that the machine was ready to operate and asked him to observe the machine in operation. The employee told an operator to turn the machine on.

The employee climbed onto the platform to see if the machine was still leaking oil. Lying on the 457-millimeter-wide platform, he observed oil on the laser resonator unit. As he was getting up, he received an electric shock. He had apparently contacted three glass tubes within the cabinet. As a result of the electric shock, he could not move his arms or speak. He scooted off the platform to the floor and ran into the office. He told them he had received an electric shock and then collapsed to the floor. The employee was treated at a hospital for burns and damage to his right index finger and left palm and for two burns to his right hip. He was released 4 hours later. That evening, while at home, the employee felt tingling and cramping and could not speak. His wife transported him to a hospital, where he was admitted.

EXAMPLE 5: ACCIDENT: 201060068—TWO EMPLOYEES
SUFFER EYE INJURIES FROM NONDIRECT LASER LIGHT

At approximately 3:00 p.m. on March 18, 1996, Employee #1, an assistant project scientist working in the laboratory of Dr. Kent Wilson at the University of California at San Diego, was aligning a 50 Hz, 800 nm tera-watt laser system for an experiment. As he was viewing the laser focus through a chamber window, he became exposed to nondirect laser light. The approximate laser intensity was 3 watts at 50 Hz, producing about 60 mJoules/pulse of energy.

During this brief period of alignment, Employee #1 had removed his protective eyewear. He suffered serious injury to both eyes that involved partial vision loss. After this incident, it was determined by the U.C.S.D. Environmental Health and Safety Department that two similar incidences had occurred earlier to another worker. Sometime on the evening of December 28, 1995, Employee #2, a research chemist in the same lab and a coworker of Employee #1, was performing the identical procedure as previously described. During the period of alignment, he removed his protective eyewear and suffered serious injury to both eyes involving partial vision loss. Employee #2 sought medical attention at a later date.

The second incident occurred on March 16, 1996. Again, alignment of the laser into the chamber was being performed. When Employee #2 removed his safety glasses to view the focus, both of his eyes were again exposed to nondirect laser light. The intensity and power of the laser were approximately the same as noted for Employee #1. The causal factors for all three incidences were unsafe work practices and, more specifically, removing protective eyewear during the alignment of the 50 Hz, 800 nm terawatt laser system. The accidents occurred at an academic research laboratory that specialized in laser science and, specifically, in the construction of high intensity, ultrashort laser and x-ray systems to view molecular dynamics.

EXAMPLE 6: ACCIDENT: 887869—EMPLOYEE'S ARM
BURNED BY CARBON DIOXIDE LASER BEAM

On February 22, 1989, Employee #1, a trained employee with 3 years of experience, was operating an Acme Cleveland Corp. model 1510-S laser cutting system to manufacture men's and women's custom-tailored suits. The system consists of a pallet shuttle that moves pallets with fabric into a cutting area with an enclosed top, bottom, and two sides. Plexiglass shielding is located at the pallet entrance and exit, with a small square opening at the lower corner on each side. The pallet is removed for review after cutting and then passed on to another portion of the shuttle, where the pallet is lowered and returned to the beginning of the system. All movement is controlled by an operator via computer console. Daily test cuts are made prior to beginning work to detect problems, and weekly and monthly maintenance checks are also performed per the manufacturer's specifications.

Documented major repairs are performed by an outside contractor. On the day of the accident, a test cut was performed and, because no problems were detected, Employee #1 inserted a pallet, initiated the first cut, and moved away to review documents in preparation for the next cut. When she returned to the control position, she noticed that no cuts had been made. She leaned forward, looked into the plexiglass, and reached up with her right hand to hit the halt button. She felt pain in her left forearm. Employee #1 had sustained second- and third-degree burns from a carbon dioxide laser beam.

She was treated at a local hospital and released to return to work after scheduling follow-up visits. Subsequent investigation by the company found that a screw had become loose on one of five reflective mirrors that controlled the laser path direction. The resulting shift in the mirror allowed the beam to overshoot it and exit the cutting area. The manufacturer has built-in provisions for automatic shutdown of the laser due to overheating, but not for a misalignment of the beam path. A maintenance technician resecured the mirror, checked all other mirrors, and applied Locktite to all screws to prevent a reoccurrence of this incident. The same was done for the employer's two other systems, which are identical to the unit involved in this accident.

EXAMPLE 7: ACCIDENT: 823294—ELECTRIC SHOCK: DIRECT CONTACT WITH ENERGIZED PARTS

A project engineer informed two maintenance workers that a Rofin-Sinar 1000-watt carbon dioxide laser was not cutting to standards. One of the maintenance workers, who was the head of the maintenance department and an experienced electrician, went with the project engineer, who showed him that the discharge current on one of the eight tetrodes was low. The electrician opened the left door at the front of high-voltage cabinet and pulled out the drawer containing four of the eight tetrodes. He then began adjusting the discharge current using a small pocket screwdriver. The laser was being used in production at the time and being operated by a production worker.

The engineer twice asked the electrician if the laser should be deenergized (once before the drawer was opened). The electrician, who had received his training from the laser manufacturer, said not to. The engineer walked away after the electrician completed the adjustments to the left side of the high-voltage cabinet. The electrician then opened the right door, opened the drawer containing the other four tetrodes, and began adjusting them. The second maintenance worker came over, and the two employees discussed the problem with the laser. The electrician then returned to his work. His coworker informed him that his procedure was unsafe and noted that the job could be done safely with the equipment deenergized.

The electrician ignored this advice and continued to work. As the second maintenance worker turned leave, he heard and saw sparking and saw his coworker slump to the floor. The electrician had contacted energized parts operating at 21.5 to 30 kilovolts, dc, and was electrocuted. The manufacturer had equipped each of the high-voltage cabinet doors with an interlock to deenergize the enclosed circuits; however, the interlocks on both front doors had been bypassed during the adjustment and might even have been bypassed well beforehand.

EXAMPLE 8: ACCIDENT: 125986612—EMPLOYEE'S LEGS CRUSHED WHILE IN LASER CUTTING PIT

At 4:50 p.m. on January 7, 2003, Employee #1, a machine operator for tortilla-making machines, was trying to reattach a laser-cutting head to its connection point on the laser-cutting machine. The machine had a bed or a table on which a sheet metal was laid and moved along a y-axis. Beneath the bed was a pit where the cut-out pieces were to fall from the sheet. When the laser head was moving over a bumpy spot of the sheet metal, it tipped off the head from the connection.

Just before the incident, the laser head detached four times at the same spot on the sheet while the sheet was being cut. Each time, Employee #1 moved the connecting point to far end and reattached the head from the side of the machine. But as the problem persisted, he climbed into the pit and reattached the head so the machine could bypass the trouble spot. After reattaching the head the fourth time, he told the operator of the next shift to start the button to resume cutting. Without knowing that Employee #1 was in the pit, the new shift employee pressed the "start" button, which caused the bed of the machine to move toward Employee #1. Employee #1's lower legs were then entrapped between the frame of the bed and a pipe in the pit. Employee #1 sustained fractures to his legs and was hospitalized.

EXAMPLE 9 ACCIDENT: 201636008—EMPLOYEE'S ARM IS AMPUTATED IN LASER CUTTER

At approximately 1:45 p.m. on June 1, 2006, Employee #1 was removing a piece of metal from a laser cutting table. He reached over the table as the cutter proceeded through its cycle. The laser cutting head amputated Employee #1's left arm. He was hospitalized for treatment.

FEDERAL OSHA LASER JUSTIFICATION

If one reviews the Code of Federal Regulation that applies to OSHA, you will not find much on lasers, outside of construction lasers. What OSHA inspectors use as their regulatory basis to inspect laser use is the General Duty Clause; which states:

a. Each employer
   1. Shall furnish to each of his employees employment and a place of employment which are free from recognized hazards that are causing or are likely to cause death or serious physical harm to his employees; 29 USC 654.
   2. Shall comply with occupational safety and health standards promulgated under this Act.
b. Each employee shall comply with occupational safety and health standards and all rules, regulations, and orders issued pursuant to this Act which are applicable to his own actions and conduct.

OSHA has set certain limitations on the use of the general duty clause.

1. Section 5(a)(1) violations cannot be grouped together, but may be grouped with a related violation of a specific standard.
2. Section 5(a)(1) cannot be used to impose a stricter requirement than that required by a standard.

3. Section 5(a)(1) cannot be used to require an abatement method not set forth in a specific standard.
4. Section 5(a)(1) cannot be used to enforce "should" standards.
5. Section 5(a)(1) cannot be used to cover categories of hazards exempted by a standard.

WHAT DOES OSHA USE FOR LASER COMPLIANCE? CONSENSUS STANDARDS IS THE ANSWER

Section 6(a) of the Williams-Steiger Occupational Safety and Health Act of 1970 (84 Stat. 1593) provides that:

> ...without regard to chapter 5 of title 5, United States Code, or to the other subsections of this section, the Secretary shall, as soon as practicable during the period beginning with the effective date of this Act and ending 2 years after such date, by rule promulgate as an occupational safety or health standard any national consensus standard, and any established Federal standard, unless he determines that the promulgation of such a standard would not result in improved safety or health for specifically designated employees.

The legislative purpose of this provision is to establish, as rapidly as possible and without regard to the rule-making provisions of the Administrative Procedure Act, standards with which industries are generally familiar, and on whose adoption interested and affected persons have already had an opportunity to express their views. Such standards are either (1) national consensus standards on whose adoption affected persons have reached substantial agreement, or (2) Federal standards already established by Federal statutes or regulations.

> **1910.1(b)** This part carries out the directive to the Secretary of Labor under section 6(a) of the Act. It contains occupational safety and health standards which have been found to be national consensus standards or established Federal standards.

OSHA has used consensus standards extensively as a basis for its mandatory safety and health standards since the earliest days of the Occupational Safety and Health Act of 1970 (OSH Act), 29 U.S.C. 651. Congress provided this authority so that OSHA would have a mechanism to begin immediately protecting the Nation's workers through mandatory standards. Using Section 6(a), the Agency adopted many consensus standards as OSHA standards. OSHA adopted some of the consensus standards through "incorporation by reference." When it incorporates a consensus standard by reference, OSHA requires employers to follow a consensus standard identified by name and date in the Code of Federal Regulations. The ANSI Z136.1 Safe Use of Laser is an OSHA recognized consensus standard for laser safety in the workplace.

As a reminder, Z136.1 in section 1.2 states that the user of Z136.1 can apply the controls of other Z136 series laser standards if they are more relevant to a particular application, even if they conflict with Z136.1.

If OSHA references a standard it, as a matter of practice, does not say the latest version. Which of course can cause a problem when one follows the latest version not the older one OSHA may have cited.

OSHA has a policy of issuing "de minimis" notices to employers who comply with more current versions of consensus standards, to the extent that the more current versions are at least as protective as the older versions. OSHA inspectors would like institutions to do a comparison and evaluation of a new versus older version of such standards.

STATE OSHA

Those who deal with OSHA know it is not monolithic. A number of States have agreements with Federal OSHA to run their own "State" OSHA programs. State Plans must set workplace safety and health standards that are "at least as effective as" OSHA standards. Many State Plans adopt standards identical to OSHA. State Plans have the option to promulgate standards covering hazards not addressed by OSHA standards, such as lasers. If the State has its own agency that regulates laser, State OSHA will bow to that agency's jurisdiction. A State Plan must conduct inspections to enforce its standards, cover state and local government workers, and operate occupational safety and health training and education programs.

The following 22 states or territories have OSHA-approved State Plans that cover both private, and State and local government workers:

- Alaska
- Arizona
- California
- Hawaii
- Indiana
- Iowa
- Kentucky
- Maryland
- Michigan
- Minnesota
- Nevada
- New Mexico
- North Carolina
- Oregon
- Puerto Rico
- South Carolina
- Tennessee
- Utah
- Vermont
- Virginia
- Washington
- Wyoming

Workers at state and local government agencies are not covered by OSHA, but have OSH Act protections if they work in states that have an OSHA-approved State Plan. OSHA rules also permit states and territories to develop plans that cover state and local government workers only. In these cases, private sector workers and employers remain under federal OSHA jurisdiction.

Five additional states and one U.S. territory (Virgin Islands) have OSHA-approved State Plans that cover state and local government workers only:

- Connecticut
- Illinois
- Maine
- New Jersey
- New York
- Virgin Islands

SUMMATION

Regulatory agencies want to know when things go wrong. The pressure on the medical user is extreme to notify the appropriate agency. In real life, nonmedical reporting has not been as robust or straightforward. Which is an error we all should work to correct. It is not so much the fear of a fine, but negative publicity that prevents Universities and other institutions from reporting or reporting properly.

13 Visual Interference Hazards of Laser Light

Patrick Murphy

CONTENTS

Unexpected bright light aimed at a person can distract them, disrupt their focus, disorient them, and/or visually incapacitate them by causing glare or flashblindness. This becomes a safety hazard when the bright light interferes with a person doing a demanding task such as driving a car or flying an aircraft.

In almost all reports of bright light visual interference, the light is known or is assumed to involve lasers. This is due to lasers' low-divergence beams, relatively low cost, portability/ease of use, and widespread availability.

This chapter will primarily concentrate on the hazards of eye-safe but too-bright laser beams for pilots. It is theoretically possible for handheld lasers aimed from the ground to cause eye injury, but this is less of a concern and will be discussed in the next chapter.

Operators of other vehicles such as automobiles, trains, and watercraft have been targeted by persons misusing lasers. But it is pilot targeting which gets the most attention and causes the most concern, so that will be discussed here as a "worst case scenario."

The chapter has been written to be useful both for laser safety experts as well as for nonexperts such as pilots, regulators, reporters, and others trying to understand laser/aviation safety.

BACKGROUND

Pilot reports of laser light interference first occurred in the 1990s. A series of high-profile incidents in Las Vegas in the mid-1990s was the first indication to federal agencies and laser users that pilot vision and performance could be seriously affected by eye-safe laser illuminations. Laser shows in the Las Vegas area were suspended in December 1995.

The SAE G10-T Laser Safety Hazards committee, composed of various interested parties, had already been studying the issue in depth. Their recommendations became enacted by the Federal Aviation Administration (FAA) to help evaluate laser usage around airports. This work helped reduce the number of incidents caused by responsible laser users.

However, around the turn of the century, low-cost, higher-powered green lasers became more available to the general public and the number of incidents of ordinary people aiming lasers at aircraft steadily rose.

To help document the rise, since 2004, pilots flying in the United States have been required by FAA to report laser illumination events.* This includes when laser beams are aimed toward the aircraft, as well as reporting whether the laser light actually enters the cockpit windows or goes directly into a crewmember's eye.[†]

EXAMPLE PILOT REPORTS OF EYE EFFECT

In 2016, there were 7442 laser illumination events reported to the FAA. In 24 (0.32%) of these events, one or more pilots reported an eye effect or eye injury. Here are some representative reports:

1. At 400 feet AGL [above ground level] a green laser beam directly contacted the pilot's eyes 2 or 3 times. The duration was "extremely brief." The pilot advised that it was similar to being temporarily blinded by looking at the sun. He did a go-around and did not land at the airport. His vision has since recovered.
2. Strong but brief blue laser illumination. Report possible eye injury to pilot. Will update when more is known.
3. Laser from the south. Pilot struck in left eye. Experienced momentary double vision, but appeared to be recovered.
4. Green laser illumination event approximately 1 mile east. Pilot reported pain in his left eye.

---

* This is required by FAA Advisory Circular 70-2A, "Reporting of Laser Illumination of Aircraft." FAA also has a reporting web page, "Report a Laser Incident" at https://www.faa.gov/aircraft/safety/report/laserinfo. There are online and downloadable Laser Beam Exposure Questionnaires at this site.
† There is essentially no concern or tracking of *passengers* seeing laser light. This is because the risk of eye injury is almost nil, and any visual interference effects on a passenger would not affect flight safety.

5. Flashing green laser from the south. Captain reported injury to right eye. Advised ATC he was seeking medical attention.
6. Illuminated by a very powerful green laser. Pilot was incapacitated. He was seeing dots and could not adjust his vision even after 10 minutes.
7. Student pilot was disabled from green laser illumination as he departed. Instructor had to take over control of the aircraft.
8. Green laser illumination lasting roughly 2.5 minutes about 5 miles from the aircraft. The first officer reported a lasting glare in his eyes when he closed his eyelids after the event. He advised he would be seeking medical attention.

NUMBER OF LASER EYE EFFECTS AND INJURIES

Table 13.1, from LaserPointerSafety.com, shows the number of incidents where FAA recorded eye effects or injuries, for 4 selected years.

In about 6% of cases, there was no actual eye effect or injury, so the 4-year total is 160 actual eye incidents out of 22,218. This is 0.72% of all reported incidents for the 4 selected years.

One interesting statistic gleaned from the above is that eye injuries are decreasing as a percentage of all incidents. In 2011, 1.4% of incidents had reported eye effects or injuries; by 2016, this dropped to 0.3%. Perhaps pilots are better able to take actions reducing the light intensity (not looking at the beam, blocking the beam, turning the aircraft, and so on).

The total number of reported aircraft laser illuminations in five major countries from 2004 to 2016, exceeds 50,000.* Assuming that the 0.72% rate is also valid for

TABLE 13.1
Eye Injuries or Effects Reported to U.S. FAA 2011–2015

Incidents with Eye Effects or Injuries	2011	2012	2015	2015	4-Year Total
Total number of incidents	3591	3482	7703	7442	22,218
Number of incidents listed in FAA reports as having one or more eye effects or injuries	55	35	55	25	170
No effect or no injury (although listed in the "Injury" column in FAA reports)	3	4	2	1	10
Number of incidents that actually had one or more eye effects or injuries	52	**31**	53	24	160
Percent of incidents that actually had one or more eye effects or injuries	1.4%	0.9%	0.7%	0.3%	0.7%

Note: FAA did not provide LaserPointerSafety.com with detailed injury reports for 2013 and 2014.

* United States 2004–2016: 36,736. United Kingdom 2009–2014: 8565. Australia 2007–2015: 3713. Italy 2010–2014: 3124. Canada 2008–2015: 2585. Total: 54,723. Note that only the U.S. data covers the entire period mentioned in the text of 2004–2016. This means the actual number of laser illuminations in these five countries, over 2004–2016 would be greater. Data and links to original sources at http://www.laserpointersafety.com/latest-stats/latest-stats.html

other countries, that would mean there have been roughly 400 or more incidents with reported eye effects or eye injuries in those five countries, over those 12 years (about 33 per year).

Fortunately, none of these 50,000+ illuminations have caused an accident as of September 2017. There have been no bodily injuries or deaths, and no airframe damage. And as discussed in the next chapter, there are no documented or proven cases of permanent eye injuries to civilian pilots caused by lasers aimed from the ground toward the aircraft.

NATURE OF REPORTED LASER EYE EFFECTS AND INJURIES

For the 160 FAA-reported cases of actual eye effects and eye injuries that occurred in the 4 selected years, the table below shows the effects as recorded by the FAA. The total adds up to more than 160, due to some incidents causing multiple effects.

A few of the reported effects did not affect the eye; for example, headache or distraction/disorientation. But they are included in Table 13.2 with the "eye effects and injuries" since they are listed this way in the FAA database—and since they clearly can have an adverse effect on a pilot's performance.

In about 20% of cases where a pilot had eye effects or injuries, the pilot sought or considered seeking medical attention.

The FAA's Laser Beam Exposure Questionnaire* lists these adverse vision effects:

Glare (could not see past the light while it was in your eye(s))
Temporary flash blindness and/or after images (similar to a camera flash)
One or more blind spots (spots in visual field lasting longer than 5–10 minutes)
Blurry vision
Significant loss of night vision
Other (specify) _____

It then gives these definitions:

Glare: A temporary disruption in vision caused by the presence of a bright light (such as an oncoming car's headlights) within an individual's field of vision. Glare lasts only as long as the bright light is actually present within the individual's field of vision.

Flash blindness: A temporary visual interference effect that persists after the source of the illumination has ceased, similar to a bright camera flash.

After image: An image that remains in the visual field after an exposure to a bright light.

Blind spot: A temporary or permanent loss of vision of part of the visual field. Unlike an after image, a blind spot does not fade, or fades very slowly (taking many minutes, hours or days to fade out).

---

* FAA, *Laser Beam Exposure Questionnaire*, https://www.faa.gov/aircraft/safety/report/laserinfo/media/FAA_Laser_Beam_Exposure_Questionnaire.pdf

TABLE 13.2

Eye Injuries or Effects Reported to U.S. FAA for 4 Selected Years

Result of Laser Illumination	2011	2012	2015	2016	4-Year Total	% of Total Effects/Injuries
Claimed retinal damage	1				**1**	0.5%
Flashblinded and/or afterimage	10	21	7	6	**44**	22.7%
Eye injury (unspecified)	9		8	3	**20**	10.3%
Pain, burning or irritation in eye	17	9	12	3	**41**	21.1%
Blind spOt(S)	5	1	2	3	**11**	5.7%
Glare			4	5	**9**	4.6%
Loss of night vision	1		2	2	**5**	2.6%
Eye or vision impairment (other than loss of right vision)	8	1	5	1	**15**	7.7%
Blurriness	7	2	14	1	**24**	12.4%
Headache	3	3	1	1	**8**	4.1%
Momentary double vision				1	**1**	0.5%
Watering eye				1	**1**	0.5%
"Affected, feels weird"			3		**3**	1.5%
Dizziness			1		**1**	0.5%
Incapacitated				2	**2**	1.0%
Distracted or disoriented	1		6		**7**	3.6%
Grounded or similar long-term	1				**1**	0.5%
TOTAL—Number of eye effects or Injuries reported	**63**	**37**	**65**	**29**	**194**	**100.0%**
Medical attention	**2011**	**2012**	**2015**	**2016**	**4-year total**	**% of actual Incidents with effects/ injuries**
Number of persons who sought or considered seeking medical attention	7	4	17	4	**32**	20.0%

Note: FAA did not provide LaserPointerSafety.com with detailed injury reports for 2013 and 2014.

The Questionnaire also lists physical effects which pilots can report:

Watering eye(s)
Eye(s) discomfort or pain
Headache
Feeling of shock
Disorientation or dizziness
Other (specify) _____

CRITICAL PHASE OF FLIGHT

When an aircraft is in cruise at high altitude, there is plenty of time for a pilot to recover from a laser illumination. From a visual interference standpoint, this is not

Aircraft reporting laser illuminations to FAA, 2013, by altitude
From 1000 to 40,000 feet above ground level, every 1000 feet

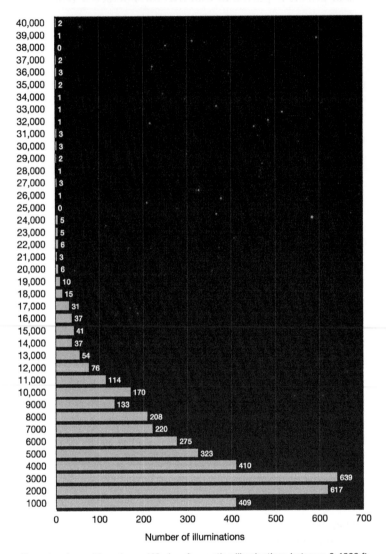

Number of illuminations

*Example: Lowest bar shows 409 aircraft reporting illuminations between 0-1000 ft.,
next bar above shows 617 aircraft reporting illuminations between 1001-2000 ft., etc.*

FIGURE 13.1 Aircraft reporting laser illuminations to FAA, 2013, by altitude.

a critical phase of flight. However, most illuminations occur at lower altitude, as
Figure 13.1 indicates:

Safety experts worry about laser events that occur during a **critical phase of
flight**: takeoff, landing, low-level maneuvering, or an emergency.

The illumination by itself is manageable, as shown by the 50,000+ laser incidents with no accidents. However, when combined with other in-flight anomalies or emergencies, laser illumination at the wrong time could be "the straw that broke the camel's back."

An article in the *Los Angeles Times* describes what laser visual interference during landing can be like.

> The beam followed the airliner for about 45 seconds before he regained enough low-light vison to see the landing area, [Captain Robert Hamilton] said. "You shouldn't rub your eyes because it makes it worse, and you can't close your eyes because you'll miss a lot traveling at 150 miles per hour. ... So, there's not a whole bunch you can do but try to use the airplane automation and try to work through that burning."*

CORNEAL DAMAGE FROM LASER IS UNLIKELY

Captain Hamilton mentioned not rubbing the eyes. The reason this is discouraged by FAA and safety experts is because rubbing can irritate the cornea, which is quite painful. Some reports of laser incidents, where pilots have sought medical attention, have said "the laser caused corneal damage."

However, it would be impossible[†] for a visible laser beam to damage the clear cornea—the light goes right through. All known reports of corneal damage were likely caused by too-vigorous rubbing of the pilot's eyes. While this could be considered a result of the entire laser illumination event, the corneal damage was not caused by the laser light—plus, the damage was preventable if the pilot knew not to rub their eyes.

For visible light, it is absorption by the retina—thermal damage—that could cause injury. The potential for such injury, along with a case of alleged corneal damage, is discussed in Chapter 14.

LIGHT LEVELS WHICH CAUSE VISUAL INTERFERENCE

Not all laser light in airspace poses a risk to flight operations. For example, in May 2005, NORAD began using a laser Visual Warning System in the Washington, DC Air Defense Identification Zone (ADIZ). An eye-safe laser beam that flashes red-green-red and is visible up to 25 miles away, is aimed directly at aircraft in the ADIZ when the pilot cannot be reached by radio.

The primary concern of laser/aviation experts is when unauthorized laser light interferes with pilots' vision during a critical phase of flight, or otherwise adversely

---

* Dave Paresh, "Pilot Says Laser Pointers Make 'Mayhem Ensue' in Cockpit," *Los Angeles Times*, February 13, 2014, http://articles.latimes.com/2014/feb/13/nation/la-na-nn-pilot-laser-pointer-airplane-20140213.
† "Impossible" at the relatively low irradiance, and short time-of-exposure of a ground-to-air laser event involving visible light. Of course, nonvisible light such as infrared or ultraviolet could in theory cause corneal damage but then again, the light would not cause visual interference and would probably be unnoticed.

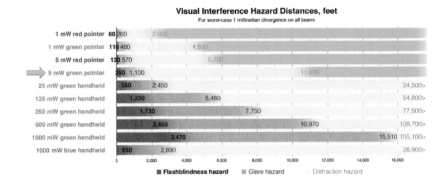

FIGURE 13.2   Visual Interference Hazard Distances.

affects their ability to safely operate the aircraft. The three main visual interference hazards that have been identified are*:

Temporary flashblindness: 100 microwatts/cm² or more
Glare: 5 microwatts/cm²–100 microwatts/cm²
Distraction†: 50 nanowatts/cm²–5 microwatts/cm²

Below 50 nanowatts/cm², the laser light is no brighter than city and airport lights as seen from the air at night. For this reason, the laser light is no longer considered to be a distraction.

Figure 13.2 shows the distances over which lasers can cause flashblindness, glare, and distraction. For example, the arrow points to the visual interference hazard distances of a 5-milliwatt green laser pointer. (This is the most powerful laser that can legally be sold as a "pointer" under U.S. Food and Drug Administration regulations.)

A 5 mW green laser pointer is a flashblindness hazard from 0 to 250 feet (red zone), a glare hazard from 250 to 1100 feet (orange zone), and a distraction from 1100 to 11,000 feet (yellow zone). Beyond 11,000 feet (green zone), it is not a visual hazard; the FAA has determined that the laser's light would be no brighter than other city and navigation lights visible at night.

Note that Figure 13.2 shows the various hazards as shading from one level into another. This is because there is no single point at which an illumination changes from "flash blinding" to "glare"—it is a gradual process.

---

* These levels were developed in the late 1990s by the SAE G10-T Laser Safety Hazards committee. The levels were arrived at in part by testing and in part by discussion and compromise. The levels were later incorporated into FAA guidelines such as Advisory Circular 70-1. Recent work by Williamson and McLin have more accurately established parameters for dazzle and glare. See for example Craig A. Williamson and Leon N. McLin, "Nominal ocular dazzle distance (NODD)," *Appl. Opt.* 54 (2015): 1564–1572. For purposes of pilot protection, the SAE-established limits are probably accurate enough.
† Distraction is not really a *visual* hazard, but is mental task disruption. A pilot who is aware of laser hazards can mentally overcome distraction. For convenience in discussion and enforcement, safety experts such as the SAE G10-OL Operational Lasers committee have grouped distraction in with the other two visual interference levels, glare and flashblindness.

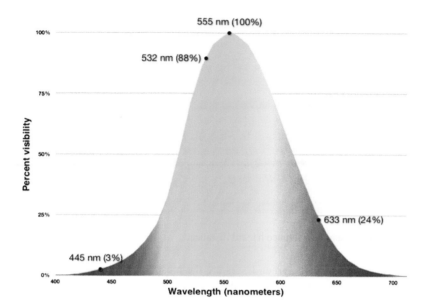

FIGURE 13.3   Perception visibility by wavelengths.

Also note that all lasers in this chart assume a 1-milliradian divergence, to keep an "apples to apples" comparison. In real life, the more powerful the laser, the higher the divergence. So, the 500 and 1000 mW lasers shown here would be expected to have higher divergence—around 1.5–2 mrad—which results in their actual hazard distances being 1.5–2 times shorter than Figure 13.2 indicates.

COLOR AS A VISUAL INTERFERENCE FACTOR

One of the key concepts the hazard distance chart illustrates is that the beam's color is an important factor in determining visual interference. That's because the human eye is most sensitive to green light. The curve in Figure 13.3 shows the eye's response to color.* A 532 nm green laser beam—the most common type sold to consumers—will appear much brighter than a 633 nm red beam or a 455 nm blue beam of equivalent power.

Another way to consider this is that a green laser beam will be a visual interference hazard over a longer distance than red or blue beams of equivalent power. Figure 13.4 shows a highlighted example. A 5 mW red laser pointer can cause glare up to 570 feet away, while an otherwise identical 5 mW green pointer can cause glare up to 1100 feet. Again, this is because the human eye is more sensitive to green light.

---

* This is the photopic response, when the eye is light adapted. The SAE G10-T determined that pilots in a nighttime cockpit have color discrimination (to see illuminated cockpit instruments) and thus the photopic curve is more relevant than the scotopic (dark-adapted) curve. The FAA uses this curve in *Advisory Circular 70-1 Table 5, Visual Correction Factor for Visible Lasers,* to determine how the light's wavelength affects visual interference distances.

Visual Interference Hazard Distances, feet
For worst-case 1 milliradian divergence on all lasers

1 mW red pointer	60 260	2,600
1 mW green pointer	110 480	4,800
5 mW red pointer	130 570	5,700
5 mW green pointer	250 1,100	11,000
25 mW green handheld	550 2,450	24,800>
125 mW green handheld	1,230 5,480	54,800>
250 mW green handheld	1,730 7,760	77,500>
500 mW green handheld	2,450 10,970	109,700>
1000 mW green handheld	3,470 15,510 155,100>	
1000 mW blue handheld	650 2,890	28,900>

0 2,000 4,000 6,000 8,000 10,000 12,000 14,000 16,000

■ Flashblindness hazard   ■ Glare hazard   Distraction hazard

FIGURE 13.4   Visual interference hazard distance.

Unfortunately for pilots, green lasers are the most common and have the longest visual interference distances, compared to otherwise identical red or blue lasers. This explains why about 95% of incidents reported to the FAA in recent years involve green laser light (Figure 13.5).

FDA PROPOSES BANNING GREEN AND BLUE LASER POINTERS

In the United States, the Food and Drug Administration (FDA) regulates laser devices. The FDA recognizes that laser/aircraft incidents were rare back when laser pointers were primarily red. According to their data, which is based on a scotopic (fully dark-adapted) luminous efficiency curve, 615 nm (red) light appears only 1.4%

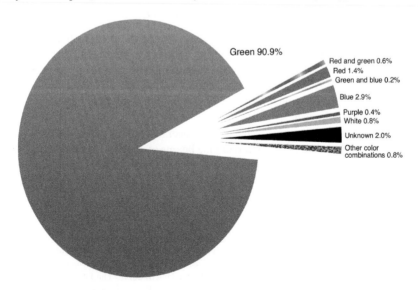

Green 90.9%

Red and green 0.6%
Red 1.4%
Green and blue 0.2%

Blue 2.9%

Purple 0.4%
White 0.8%

Unknown 2.0%

Other color
combinations 0.8%

FIGURE 13.5   Color of laser illuminations reported to FAA, 2016.

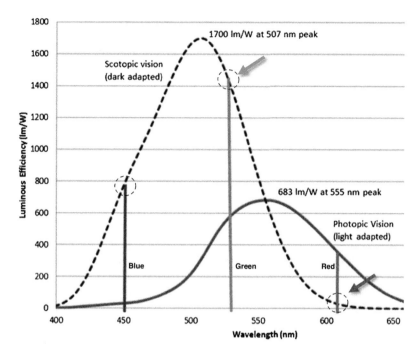

FIGURE 13.6   Luminous efficiency.

as bright as the common 532 nm light used in most current (2017) green laser pointers (Figure 13.6).*

In October 2016, the FDA proposed a novel approach to regulating laser pointers. They are considering designating lasers that emit light between 400 and 609 nanometers as "defective." This would be done under a federal regulation defining a defective product as one which "emits electronic product radiation unnecessary to the accomplishment of its primary purpose which creates a risk of injury." The only laser pointers which would be allowed to be manufactured[†] would have beams in the orange-red to deep red region of 610–710 nm.

By doing this, the FDA would make it easier for customs officers. Instead of having to measure the power of laser pointers in a shipment, they can simply allow or ban a shipment based on the beam color of the pointer. This would also make it simple for any state or locality that would like to restrict sales, possession, or use of laser pointers based on the beam color.

---

* This differs from the FAA Visual Correction Factor curve presented earlier. Using the FAA curve, 610 nm light appears about 50% as bright as 532 nm light, to pilots. The FAA's curve recognizes that pilots flying at night are not completely dark-adapted due to viewing cockpit displays and lights, so the curve is closer to photopic. It is likely that a key topic in discussion of the FDA proposal will be whether the FAA or FDA curves are more representative of pilot vision.

† Except for variances, which can be granted to manufacturers showing a need to produce a non-red and/ or higher-power laser pointer than would normally be allowed. For example, such a laser pointer might be sold only to military or law enforcement.

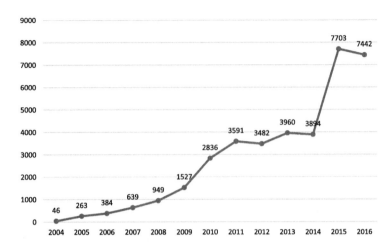

FIGURE 13.7 Laser illuminations reported to U.S. FAA, annual total.

As of September 2017, this proposal is still in a draft form. It would need to be published in the Federal Register for a public comment period. Based on the responses, the FDA could issue a final rule with the original or a modified version of the proposal, or the FDA could withdraw the proposal.*

Even if non-red laser pointers are eventually prohibited from manufacture, there still would be millions of already-existing green laser pointers in the hands of consumers. It would not be illegal, under the FDA proposal, for these consumers to possess or use green pointers. Thus, it still could take years for FDA color-based restrictions to significantly reduce the number and severity of laser illumination incidents.

NUMBER OF ILLUMINATION INCIDENTS IN SELECTED COUNTRIES

Figures 13.7 to 13.11 consist of a series of graphs showing laser illumination events in five selected countries.†

There is no known or clear reason for the United States' jump—almost doubling—from 2014 to 2015. One possibility is that increased publicity from an FBI effort in 2014 caused a "copycat" effect. But there was increased publicity in previous years as well, which did not appear to translate into significantly higher illuminations. Also, there was no significantly new or improved technology. Laser pointers continued to drop in price, with higher available powers, as they have throughout the 2000s and 2010s. However, this was part of a gradual, continuous process with no apparent discontinuity in 2015.

---

* A fuller description of the FDA proposal, its limits and implications, can be found in Patrick Murphy and Daniel Hewett, "FDA's Proposed Change to the Regulation of Laser Pointers," *Proceedings of the 2017 International Laser Safety Conference.*

† These graphs are the source of footnote 3 stating that there were at least 54,723 laser illuminations in these five countries, from 2004 through 2016.

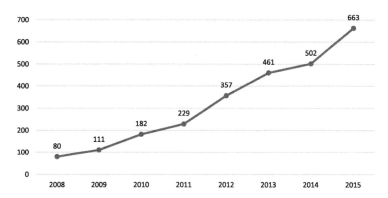

FIGURE 13.8 Laser illuminations in Transport Canada CADORS database, annual total.

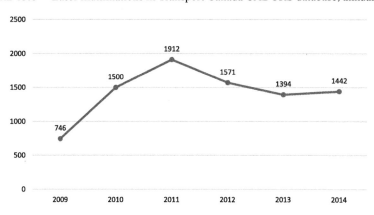

FIGURE 13.9 Laser illuminations reported to U.K. CAA, annual total.

REDUCTION OF THE HAZARD

There are a number of ways to reduce the hazard of laser visual interference with pilots (and by extension, with others doing critical tasks such as drivers). These include reducing the number of incidents, reducing the light brightness, and better education of pilots.

Reduce the Number of Incidents

The number of incidents could theoretically be reduced by methods including: educating users of the potential hazards, prosecuting and publicizing those arrested for lasing aircraft, and restricting access to handheld lasers.

The first two, education and publicity, do not seem to have worked, at least in the United States.

For a few years from 2011 to 2014, the number of incidents reported to the FAA plateaued around 3500–4000 per year. But then starting in 2015, the number nearly doubled, and has stayed that way. Unfortunately, there does not seem to be any specific cause for this doubling; there were no high-profile incidents, and no new technologies or other new factors around 2015.

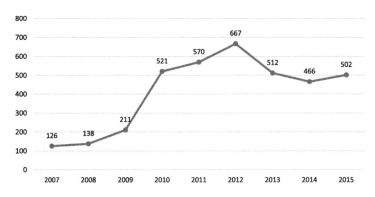

FIGURE 13.10   Laser illuminations reported to Airservices Australia, annual total.

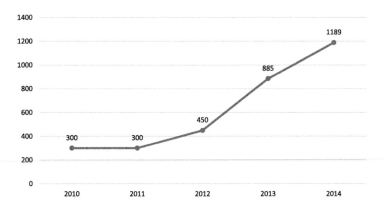

FIGURE 13.11   Laser illuminations reported to Italian ANSV, annual total.

One possibility is that an FBI publicity campaign started in early 2014 could have triggered "copycats," or otherwise influenced persons who previously had not considered aiming at aircraft. But there had also been previous publicity campaigns, with no significant effect on incident rates.

Restricting access to lasers is also difficult. The FDA does not currently have the power to restrict laser usage based on factors such as being handheld. Any new proposal, such as their October 2016 proposal to ban non-red laser pointers, would need time for comments and implementation. Millions of pre-ban handheld lasers would remain on the market.

There may also be unintended consequences. After Australian states banned higher power laser pointers in 2008, importers brought in such lasers anyway with labels that said they were the lower, legal, 1 mW pointers. This created a new hazard—lasers which claimed to be safe but actually were far over safe limits; by as much as 50–125 mW.

And despite the 2008 ban, the number of incidents in Australia continued to climb in the 4 years after the ban. They increased from 138 incidents pre-ban in 2008, to 667 post-ban in 2012.

In the United States, similar steps such as a new federal law in 2012 making it illegal to aim at aircraft, or increased prosecutions of existing laws, do not appear to have significantly affected the incident rate.

Laser Glare Protection Eyewear

Some pilots are using Laser Glare Protection (LGP) eyewear. This is specially designed to reduce (not eliminate) the most common wavelength(s) of laser light. It also is designed so cockpit instruments and airport lighting colors can still be adequately distinguished.

If a pilot is considering using Laser Glare Protection, here are some tips:

1. **First responders should consider LGP eyewear.** They have a greater chance of encountering laser illumination, and having to perform a mission despite the laser light. Usually, commercial and general aviation pilots would have a briefer exposure time and/or can maneuver to avoid the laser light.
2. **Choose from LGP eyewear designed specifically for pilots.** Laboratory eyewear and consumer "blue blockers" or standard sunglasses are not appropriate or safe.
3. **The glasses first must be safely tested,** on the ground and while at cruising altitude, on the actual equipment being used. The glasses should still allow sufficient color rendition and sufficient color discrimination of cockpit instrumentation and airport lights.
4. **LGP glasses should NOT be routinely worn.** Instead, keep them readily available during flight so they can be deployed and worn in the event of a laser illumination or if laser activity has been reported in the area.
5. **Glasses that attenuate 532 nm green laser light should be sufficient** in most cases. Over 90% of FAA-reported incidents involve green laser light at 532 nanometers.
6. Glasses that attenuate two or more wavelengths, such as green-plus-red or green-plus-blue, are available. However, the more wavelengths that are attenuated, the greater the **possibility of reduced color discrimination of cockpit instruments.**
7. **It is not necessary for the glasses to fully block the laser's light.** Simply attenuating the light by one or two orders of magnitude (Optical Density 1 or 2) can be sufficient to prevent flashblindness and reduce glare. Informal tests have shown that for single-wavelength green (532 nm) protection, OD 3 blocks too much light (both laser and instrument panel); for example, it is not possible to track the laser beam to its source. Thus, **for green protection an OD of 2 or 2.5 is recommended**. For attenuating red or blue, a lower OD such as 1 or 1.5 is suggested. As stated above, testing on the actual equipment to be flown is essential.

Windscreen Protection

As of September 2017, two companies are independently developing a Laser Glare Protection film to be applied to the inside of cockpit windscreens. The film reflects unwanted wavelengths, so the laser light appears much dimmer and is manageable inside the cockpit. This has no effect on in-the-cockpit color discrimination since the pilot does not need to wear anti-laser eyewear.

Film from one of the companies, Metamaterials Technologies Inc., has been tested by Airbus and may go into service around 2018. It can be applied both in new aircraft, and can be retrofitted during "B check" servicing in a hangar. (Film from the other company, BAE Systems, is in the laboratory R&D testing stage.)

One disadvantage is the potential cost of putting film on the various windows in a cockpit. The advantage of course is that the pilot is automatically protected, at least from the wavelengths the film is designed to reflect.

Also, deployment would be up to individual aircraft operators. As of 2018, there are no national aviation authorities considering making such film mandatory.

DISADVANTAGES OF LASER GLARE PROTECTION

Laser Glare Protection in general can only protect against one or a few wavelengths (colors). The more colors added—especially if they are close together such as 520 and 532 green—the lower the overall light transmittance and the greater potential for color discrimination problems.

As of mid-2017, the most common color, by far, in laser/aircraft incidents is green. This is currently produced by lasers emitting 532 nm light. Most LGP protects against 532 nm light, and perhaps a few other colors such as red or blue.

But new, diode-based laser pointers which emit 520 nm light are on the horizon. When they become common, the older 532-based LGP will not be effective. And if LGP is formulated to reject both colors, there is more likelihood that the LGP will interfere with overall green color discrimination. (This may not be as much of a problem for windscreen-applied LGP as it is for LGP eyewear.)

REDUCE THE LIKELIHOOD OF ADVERSE REACTIONS

The final major approach to reducing laser aviation hazards is pilot education.

In its most simplified form, pilot education can consist of reading material and videos explaining the laser hazard. It can emphasize to pilots that this is "just bright light" with no impact or eye injury potential. It can give tips and procedures to follow if exposed to laser light.

Even better is letting pilots experience what it is like to have bright light interference while flying—in a simulator, of course.

A 2003, FAA study showed that after being exposed to laser strikes in a simulator a few times, pilots were better able to handle their aircraft. The study's authors wrote: "Post-flight comments indicated that familiarization with the effects of laser exposure, instrument training, and recent flight experience in the aircraft type may be important factors in enhancing a pilot's ability to successfully cope with laser illumination at eye-safe levels of exposure…Acquainting pilots with low-level laser exposure could minimize its effects and reduce the chance of an extreme reaction."*

Testing by the SAE G10-OL committee in August 2016 showed that even using a low-cost ($20) green LED flashlight, set by eye to glare-inducing levels, was an acceptable substitute to demonstrate to pilots in a simulator cockpit what it is like

---

* "The Effects of Laser Illumination on Operational and Visual Performance of Pilots During Final Approach," DOT/FAA/AM-04/9, US DOT FAA Office of Aerospace Medicine, August 2003. Study produced by Van Nakagawara, Ronald Montgomery, Archie Dillard, Leon McLin and Capt. Bill Connor.

to experience a laser illumination. This is safe for the pilot, as it uses conventional non-coherent light at a relatively low (glare-inducing) level.

The committee's only caveat was that the LED flashlight should NOT be used to demonstrate Laser Glare Protection eyewear. This is because LEP would not block enough of the broadband light. Pilots would thus not have an accurate demonstration of how well LEP works against its targeted wavelengths.

14 The Potential for Eye Injuries from Lasers Aimed at Pilots

Patrick Murphy

CONTENTS

It is understandable that pilots and others concerned with aviation safety would wonder about the potentially injurious effect of lasers on pilots' eyes.

The short answer is that the adverse effects of visual interference—including distraction, disruption, and visual incapacitation—are considered by knowledgeable safety experts to be much more of a concern than potential eye injuries. Statistically, there is a far greater likelihood of visual interference that could lead to an incident or an accident, than of a pilot receiving a serious eye injury (whether or not the serious eye injury led to an incident or accident).

One conclusion is that in situations sometimes faced by first responders, such as having to go after a criminal or having to rescue a person, a laser-exposed pilot should continue the mission if possible (if not dazzled or flashblinded) and should not worry about retinal injury.

As long as the pilot can take reasonable steps such as not looking into the beam, and turning the aircraft away from the source, the first responder should be able to continue the critical part of their mission.*

---

* The Coast Guard in particular has had strict rules about breaking off missions if crew is illuminated by laser light. There have been some rescue missions disrupted due to this rule. In the author's opinion, military and police first responder personnel who routinely deal with dangerous situations should continue these missions instead of overly worrying about the low potential risk of serious eye injury. A list of some lasings of Coast Guard aircraft can be found online at http://www.laserpointersafety.com/news/news/aviation-incidents_files/tag-coast-guard.php#on.

NO KNOWN OR PROVEN PERMANENT INJURIES

As of September 2017, there have been no known or proven permanent injuries to civil pilots from lasers directed at aircraft.*

For example, on March 19, 2015, an FAA spokesperson said: "The FAA is unaware of any U.S. commercial pilot who has suffered permanent eye damage as a result of exposure to laser light when in the cockpit." An April 2012 report from the U.K. CAA also stated that "there have been no documented cases anywhere in the U.K. where civil aircrew have suffered permanent eye damage as a result of a [laser] attack." Since those statements, there has been only one public case claiming permanent damage—and this case is highly questionable as discussed below.

Since laser incident reports began in the early 1990s, there have been a few pilots who claimed permanent or serious injuries. In all cases known to the author, the injuries either were minor enough to heal with no permanent adverse visual effect— the pilots could resume flying—or the claims of injuries could not be independently verified by qualified medical specialists.

Three of the most high-profile claims of serious injury are discussed below.†

MID-1990s: CORNEAL BURNS CLAIM

Glendale (CA) police sergeant Steve Robertson said that in the mid-1990s, his corneas were severely burned when his police helicopter was flashed with green laser light. He said he was momentarily incapacitated and would have crashed if his copilot had not been able to land the aircraft. He was taken to a nearby hospital "where doctors scraped his corneas." He returned to work 4 days later and still had 20/20 vision as of 2010.

The idea of visible lasers causing a corneal burn is highly suspect. This is because visible laser light goes through the transparent cornea and is not absorbed to any appreciable extent. If the visible light was strong enough to damage the cornea, it certainly would have caused extensive retinal injuries as well.

In some cases, a pilot will vigorously rub their eyes, causing painful corneal abrasions. As discussed earlier, while due to a laser light illumination incident, this was not directly caused by the light itself.

Finally, this incident is unusual because in the mid-1990s, there were no powerful handheld green lasers. Such a laser would have been large, heavy, and expensive. The source could be from a laser light show—which would have been relatively easy for law enforcement to locate—or perhaps from someone misusing an industrial laser by aiming its beam outside, then reflecting it up into the air.

---

* There may have been injuries to pilots in combat zones, from deliberate laser use by hostile forces. If any such injuries occurred, reports would have been classified and the military would take appropriate action. This chapter will discuss only publicly reported cases.

† Additional stories, about less-serious injuries or about pilots who sought medical attention after an incident, are at LaserPointerSafety.com in the *Aviation-related News* page, categorized with the phrase "Eye effect or injury." All articles are available at http://www.laserpointersafety.com/news/news/aviation-incidents_files/category-eye-effect-or-injury.php#on.

1997: KAPITAN MAN/STRAIT OF JUAN DE FUCA INCIDENT

U.S. Navy Lieutenant Jack Daly claimed to have been injured on April 4, 1997, by a red laser aimed at him as he was in a Canadian Armed Forces helicopter photographing a suspected Russian spy ship, the *Kapitan Man*, on the U.S. side of the Strait of Juan de Fuca. He suffered intense pain in his right eye. A retinal specialist found three tiny retinal pigment defects in one eye and attributed it to the laser.

There are many interesting aspects to this long-lasting case, including a Coast Guard search of the ship, Daly's appearance before a Congressional committee, his lawsuit against the shipping company, and a request—turned down—for a Purple Heart. As part of these events, he had many additional medical exams over the years.

Daly's records were summarized by laser injury experts writing in an article in the August 2004 *Archives of Ophthalmology*. They concluded that "no evidence of laser injury was found in the years after the incident by 17 other ophthalmologists, including 5 neuro-ophthalmologists and 8 retina specialists.... The few tiny RPE defects on which the initial diagnosis were based are common... The patient had real complaints, but they were caused by preexisting autoimmune problems rather than by laser injury."

2015: "MILITARY-STRENGTH" LASER SAID TO INJURE U.K. PILOT

In spring 2015, an unnamed pilot in a British Airways plane landing at Heathrow Airport was illuminated by what was assumed to be a "military-strength" laser, according to the general secretary of the British Air Line Pilots' Association (BALPA). The man was treated at a Sheffield hospital for a burned retina in one eye, and has not worked since the incident. The incident was first publicly revealed by BALPA in a November 2015 press release.

In January 2016, a medical journal report was published by two ophthalmologists and a laser safety regulator. The report stated that there was no long-term negative effect on vision: "The pilot's symptoms fully resolved 2 wk later."

In April 2016, the case was discussed in an editorial written by three leading U.K. laser safety experts—including the laser safety regulator who was a coauthor of the January 2016 medical journal report. The experts concluded the case is suspect for a number of reasons and they do not believe laser targeting caused the alleged injury. They wrote: "This case is suspect because first and foremost, the metrology and exposure geometry would suggest insufficient energy could have entered the eye to produce irreversible damage and second the fundus anomaly is in the wrong location, the wrong shape and resulted in an extremely transient reported loss of VA [visual acuity] with full recovery."

REASONS FOR LOW RATES OF PILOT EYE INJURIES

There are a number of reasons why pilots have not had documented permanent or serious eye injuries, even when Class 4 lasers have been deliberately directed at them.

Beam divergence. The distance from laser source to a helicopter or aircraft cockpit can be hundreds of thousands of feet. This gives the beam room

**Less than 1% of this laser beam's power
goes into the pupil of an eye 500 feet away**

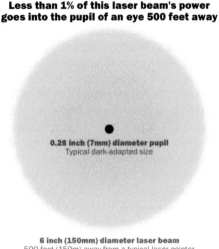

0.28 inch (7mm) diameter pupil
Typical dark-adapted size

6 inch (150mm) diameter laser beam
500 feet (150m) away from a typical laser pointer
with 1 milliradian divergence

FIGURE 14.1   Diameter of laser beam at 500 feet.

to spread, so that only a fraction of the total beam area goes through the (nominal) 7 mm aperture of an eye's pupil. For example, a Class 4 beam of 500 mW, 1 mrad at a distance of 500 feet spreads out to half foot across and has an irradiance just below the Maximum Permissible Exposure for a 0.25 second unwanted exposure time. In other words, the beam is considered unlikely to cause injury to a person who is not deliberately staring into the light.

Figure 14.1 illustrates how beam divergence reduces the power that goes into a person's pupil.

Additional safety reduction factor. Without going into too much detail, the Maximum Permissible Exposure (MPE) and the related Nominal Ocular Hazard Distance (NOHD) contain a safety or "reduction" factor. If a laser has an NOHD of 500 feet, at about one-third of that distance (167 feet) is the point at which there is a 50/50 chance of an exposure under laboratory conditions of causing the smallest medically detectable eye injury.* This means that even a nominally hazardous exposure (within the NOHD) could result in minimal or no change to the retina.

Relative movement. It is difficult to keep a handheld laser's beam steady on a target that is hundreds or thousands of feet away. In most laser/aircraft incidents, the beam is flashed briefly, one or more times, at the pilot's eyes. These short flashes give the retina time to cool down and thus reduces the

---

* "Medically detectable" meaning detectable by procedures available in the 1970s when laser retinal injury research was being used to determine safety standards. Today's medical instruments could detect smaller changes (e.g., caused by lower irradiance levels). Whether such visually imperceptible changes—imperceptible both to the person exposed to laser light, and to a trained researcher looking for changes—qualify as adverse injuries is a philosophical question.

injury risk. Also, the aircraft often is in motion—hovering helicopters excepted—also increasing the difficulty of keeping a beam on a target.

Windscreens spread the light. Aircraft windscreens often have scratches or other imperfections which scatter and/or refract the light. This is detrimental from a visual interference standpoint, as it can spread the laser light so it blocks the entire windscreen. But from an eye safety standpoint, spreading the light also reduces the irradiance.

Pilot actions (aversion effect). A person unexpectedly exposed to very bright light will reflexively blink, turn, move out of the light, or take other evasive actions. Laser safety standards take this into account. For example, many laser standards are based on a person not being exposed to visible laser light for more than quarter of a second; it is assumed the aversion response limits the exposure time. This may or may not be completely true* but the general principle of a person being safer if they turn away is a factor in helping reduce pilots' exposure time.

EVALUATING POTENTIAL LASER EYE INJURIES

In the August 2004 journal article,[†] "Assessment of Alleged Retinal Laser Injuries," the authors set forth six questions that facilitate the diagnosis of such injuries:

1. Are there ocular abnormalities that could have been caused by a known laser-tissue interaction at the time of the reported incident?
2. If the answer to 1 is "yes," have those abnormalities been documented by a reliable technique, such as fundus photography, fluorescein angiography, or optical coherence tomography?
3. If the answers to 1 and 2 are "yes," do findings from ophthalmoscopy and retinal imaging evolve after the incident in a manner consistent with a laser injury?
4. If the answer to 1 is "yes" and substantial visual or somatic complaints are present, is there any scientific evidence that the objective ocular findings could cause the reported subjective complaints?
5. If the answer to 1 is "yes" and substantial visual complaints are present, is the location of Amsler grid or visual field defects stable and consistent with the location of the retinal abnormalities supposedly responsible for causing them?
6. If the laser source involved in the alleged injury is available or known, is it capable of producing the observed clinical findings under the reported exposure conditions?

If the answers to all six questions are "yes," then a laser injury has almost certainly occurred.

---

* See BAUA, *Damit nichts ins Auge geht—Schultz vor Laserstrahlung* (Keeping an eye on safety—Protection against laser radiation), 1st ed. (Dortmund: BAUA, 2013), 32, https://www.baua.de/DE/Angebote/Publikationen/Praxis/A37.pdf?__blob=publicationFile.

[†] *Arch Ophthalmol.* Vol. 122, August 2004.

The authors state that "perceived ocular injuries with no demonstrable tissue damage are not real laser injuries."

In their concluding "Comment" section, the authors state that injuries that cause serious visual problems are readily apparent from tests, while injuries that are subtle or ambiguous "should have excellent visual prognoses and clinical outcomes."

15 Outdoor Laser Safety

Ken Barat

CONTENTS

For most laser users and laser safety officers, with the exception of the military, outdoor laser safety seems to be a rare concern. So, when one talks about laser safety outdoors we think of a "range" where the laser system is being used.*

Many outdoor laser applications are becoming more common and some are on the edge of becoming common place. Think of the Drone, Unmanned vehicle. At one time, this was the sole domain of the military and related applications. Today, drone use is not only being used for commercial applications but is a growing hobby. Research drone applications are being developed, from measuring the density of vegetation to cutting leaves on tall trees.

One other outdoor application which the public hears about indirectly is LIDAR, laser light radar—the largest application of this is driverless vehicles.

BACKGROUND MATERIAL

LIDAR, an acronym of Light Detection And Ranging (sometimes Light Imaging, Detection, And Ranging), is a surveying method that measures distance to a target by illuminating that target with a pulsed laser light, and measuring the reflected pulses

---

* While I do not know of any outdoor laser accidents that are not laser light show related or pilot illumination based, I am rather sure some have occurred or there have been near misses. All I wish to say is the modern LSO should receive some training on the topic of outdoor laser safety, if not for use at their facility, but to be aware if questions come up.

with a sensor. Differences in laser return times and wavelengths can then be used to make digital 3D-representations of the target. The name LIDAR was originally a portmanteau of light and radar.

LIDAR is popularly used to make high-resolution maps, with applications in geodesy, geomatics, archaeology, geography, geology, geomorphology, seismology, forestry, atmospheric physics, laser guidance, airborne laser swath mapping (ALSM), and laser altimetry. The technology is also used for control and navigation for some autonomous cars. LIDAR is sometimes called laser scanning and 3D scanning, with terrestrial, airborne, and mobile applications.

Mobile LIDAR (also *mobile laser scanning*) is when two or more scanners are attached to a moving vehicle to collect data along a path. These scanners are almost always paired with other kinds of equipment, including GNSS receivers and IMUs. One example application is surveying streets, where power lines, exact bridge heights, bordering trees, and so on, all need to be taken into account. Instead of collecting each of these measurements individually in the field with a tachymeter, a 3D model from a point cloud can be created where all of the measurements needed can be made, depending on the quality of the data collected. This eliminates the problem of forgetting to take a measurement, so long as the model is available, reliable, and has an appropriate level of accuracy.

LIDAR allows you to generate 3D maps, which you can then navigate the car or robot predictably within. By using a LIDAR to map and navigate an environment, you can know ahead of time the bounds of a lane, or that there is a stop sign or traffic light 500 m ahead. This kind of predictability is exactly what a technology like self-driving cars requires, and has been a big reason for the progress over the last 5 years.

OUTDOOR TECHNICAL CONSIDERATIONS

Outdoor laser work has many technical issues that typical indoor laser lab work does not encounter. Which adds to the unique hazard evaluations that are required for outdoor work.

BAD WEATHER

Weather conditions such as snow, rain, or fog generally reduce visibility and transmission. Nevertheless, in some rare conditions, reflections from raindrops or snowflakes might cause a hazard. This is not expected to cause a hazard at distances greater than 1 m from raindrops or snowflakes illuminated by the laser beam.

ATMOSPHERIC SCATTERING AND ABSORPTION

When considering distances greater than a few kilometers from the laser source, a number of factors come into play. The major ones for attenuation consists of: Mie scattering, Rayleigh scattering, and molecular absorption. Mie (or large particle) scattering occurs where the particle size is greater than the wavelength of the optical radiation, and is normally the greatest contributor in the visible and near infrared. Rayleigh (or molecular) scattering occurs where particle size is much less than the wavelength, is reasonably constant for a given wavelength, and is the greatest contributor in the ultraviolet. Except in the infrared, absorption by gas molecules is normally insignificant in comparison to scattering.

Rayleigh scattering, which is proportional to λ^{-4}, is significantly more pronounced at shorter wavelengths; thus a laser with a long wavelength (up to 2000 nm) is attenuated less than a laser that operates in short wavelengths (down to 400 nm). A clear atmosphere may, therefore, be expected to be relatively transparent to visible and near infrared wavelengths. If the effects of atmospheric attenuation are to be considered, knowing the meteorological range V (km) is necessary. The attenuation is usually based upon a V of 60 km or "Exceptionally Clear Atmosphere" since the visibility may not be known during an outdoor operation.

ATMOSPHERIC SCINTILLATION

Scintillation is a complex phenomenon that is dependent upon laser beam and atmospheric characteristics. As a laser beam passes through the atmosphere, it may undergo localized changes in irradiance or radiant exposure. "Turbulons" are created when local temperature gradients produce variations in air density perpendicular to the beam propagation direction. These "turbulons" are expressed as local variations in the air index of refraction. Passing through a portion of the atmosphere where "turbulons" are present causes different areas of the beam to focus or defocus. Scintillation effects are most prominent close to the surface of the Earth.

UPWARD DIRECTED BEAMS

Upward directed laser beams experience less atmospheric absorption and scattering due to thinning of the atmosphere; therefore, attenuation values that are used for nearly horizontal laser beams are not appropriate. For lasers aimed 30° or more above the horizon, the transmission of the entire thickness of atmosphere will typically be about 60% to 80% for a day with good visibility and no cloud cover. Lasers aimed 5° to 30° might experience some greater attenuation. Nevertheless, attenuation for these beams is greatly reduced compared with the attenuation for horizontal beams (under 5°).

SAFETY REQUIREMENTS FOR MILITARY LASER RANGES

Here are the generic range safety guidelines for military laser ranges (these are based on U.S. Navy protocols for outdoor laser work):

GENERAL

   a. Ensure that only those laser installations and ranges which have been certified by an RLSS and approved by the activity LSSO as safe for specific applications using specific laser systems are allowed to operate and then solely for those applications. Laser systems shall not be fired outside of these LSSO-designated areas and targets.

b. Range safety personnel with laser safety training and experience appropriate to the exercise or operation shall be present during all laser operations.

c. When ground positions are designating for aircraft, an aircraft exclusion zone shall be established that is centered on the ground-lasing position to the target. The exclusion zone shall be, at a minimum, a 20-degree safety cone around the firing point extending back from the target to the firing point.

d. During airborne laser operations, personnel in the lasing aircraft must wear laser protective eyewear in single aircraft laser scenarios if there is a possibility of retro-reflectors or other flat specular reflectors in the target area and within one-half the NOHD from the aircraft.

e. All personnel in other aircraft that must fly in the restricted airspace through a defined laser hazard area must have suitable laser protective eyewear in place during transit of that hazard area.

f. Class 3B and Class 4 laser target designators and rangefinders shall not be activated until a designated target has been acquired optically or through a recognized tracking system (e.g., FLIR or radar). Laser target locators and illuminators require special care during use to avoid illuminating non-target areas.

g. No Class 3B or Class 4 lasers shall be directed above the horizon unless coordinated with the Federal Aviation Administration and affected DoD components, including United States Strategic Command, JFCC Space/JOCC.

h. All ship-towed targets shall adhere to requirements of MIL-HDBK-828 series, *Laser Range Safety.*

RANGE CERTIFICATION

NSWCDD, NSWC Corona, or an RLSS shall perform complete laser radiation hazard surveys and evaluations of laser ranges to determine the degree of laser radiation hazard and to recommend proper controls. These hazard surveys and evaluations shall be performed on all new laser ranges, whenever changes to the range will exceed the current certification and every 3 years. Additionally, laser facilities and laser ranges shall receive local safety compliance inspections annually by a TLSO, LSS, or RLSS.

RANGE REGULATIONS/RANGE STANDARD OPERATING PROCEDURES (SOP)

Every laser range complex shall develop and maintain a range SOP. The SOP should contain, at a minimum, a description of the authorized firing points, run-in headings, altitude restrictions, firing fans, and other control measures and restrictions for the range.

USE OF A LASER RANGE

There are several requirements that have to be met prior to an activity or command conducting training or operations on a certified laser range.

APPROVING RANGE USE

Once the range has been certified and the RLSO has developed the SOP or range regulations, the range will be ready for use. Prior to authorizing a commands use of the range however, the requesting command will need to provide the RLSO or their designated representative with a proposed training plan. This plan will typically be reviewed by that commands ALSO to ensure compliance with the range regulations or SOP but the RLSO is still responsible to ensure there are no planned deviations from those regulations. It is the RLSO's responsibility to ensure the range regulations or SOP is available to the requesting command.

The training plan can legitimately be provided via a phone call, email, through official correspondence, or over the radio as the aircraft approach the range. Ultimately, it is up to the RLSO as to what they will accept. The idea is to ensure range safety without adding unnecessary burdens on its users. The procedures and requirements that you develop to accomplish this should also be provided within the range regulations or SOP.

In general, the training plan should include:

1. Name and date of qualification of the command LSSO.
2. Laser devices to be used.
3. Laser device firing positions.
4. Targets to be used /target areas to be used.
5. Ground personnel locations (indicating those requiring laser eye protection).
6. Laser eye protection to be used (if applicable).
7. Aircraft run-in headings (if applicable).
8. Ship heading for towed target operations (if applicable).
9. Laser mode(s)/tactics to be employed (e.g., force-on-force, designation, rangefinding, offset lasing, high altitude release bomb (HARB), and so on).
10. Hazard areas to be cleared of nonoperating personnel (roadblock locations, if required).
11. Types of surveillance to be used to ensure a clear range (if different from established procedures).
12. Radio frequencies (or channels) and standardized terminology for communication where appropriate.
    a. The RLSO for the hosting range complex shall:
       1. Ensure requesting unit has a certified LSSO coordinating the test/ training operation.
       2. Provide the local range regulations/standard operating procedures to the LSSO of the requesting unit.
       3. Review proposed laser range operations plan or test plan to ensure compliance with current certification and local regulations and standard operating procedures.
       4. Ensure a laser safety inspection of the range is completed prior to its use (e.g., signs are posted, area is clear of specular reflectors, LEP is available, and so on).
    b. The command requesting use of the laser range shall:
       1. Review host range complex range regulations/SOP.

2. Provide a range use operations plan/test plan to RLSO that includes:
   a. Name and date of qualification of the command LSSO.
   b. Laser devices to be used.
   c. Laser device firing points.
   d. Targets to be used /target areas to be used.
   e. Ground personnel locations (indicating those requiring laser eye protection).
   f. Laser eye protection to be used (if applicable).
   g. Aircraft run-in headings (if applicable).
   h. Ship heading for towed target operations.
   i. Laser mode(s)/tactics to be employed (e.g., force-on-force, designation, rangefinding, offset lasing, high altitude release bomb (HARB), and so on).
   j. Hazard areas to be cleared of nonoperating personnel (roadblock locations, if required).
   k. Types of surveillance to be used to ensure a clear range.
   l. Radio frequencies (or channels) and standardized terminology for communication where appropriate.
3. Ensure all personnel involved in operations receive an appropriate pre-mission brief to include:
   a. Authorized tactics, firing positions, firing fans, and aircraft run-in headings (as appropriate).
   b. Drawings, photographs, descriptions or grid points of authorized targets.
   c. Communication procedures that include specific frequencies (or channels), controlling authorities, and standardized terminology.
   d. Acquisition, identification, and tracking procedures for targets are established prior to laser activation.
   e. Missile/ordnance mode of operation (as appropriate for live fire operations).
   f. Requirements for beam termination.
   g. Control measures to minimize the risk of unauthorized personnel or aircraft entering the range area.
   h. Type of eye protection to be worn.
   i. Potential hazards posed by the laser system (e.g., phantom targeting and backscatter); the target area, maintenance area, and so on.; types of warning signs to be posted; and specific procedures to avoid these hazards (as appropriate).
   j. Target area, maintenance area, and so on.
   k. Warning signs.
   l. Other specific procedures.
4. Ensure appropriate laser eye protection is provided to all personnel within the laser hazard zone.
5. Ensure all aspects of the range regulations/standard operating procedures are adhered to during the operation/exercise/test.

6. Ensure only tactics, authorized within the scope of the range certification, and only LSRB-approved laser systems are used for the operation/exercise/test.

CRITICAL ELEMENTS OF AN OUTDOOR LASER SAFETY PLAN

SCOPE, PURPOSE, RATIONALE, AND DURATION

- Include specific project and field description
- Follow internal laser safety standards
- Describe duties of Laser Safety Personnel
- Laser Description or specifications
- Eye Hazard/Safety Calculations
- Follow ANSI standard Z-136.1 "Safe Use of Lasers" and ANSI Z136.6 "Safe Use of Lasers Outdoors"
- Test Area Access Control Requirements
- List Laser Operational Restrictions
- Weather Safety Test Layout
- Test Safety Precautions (Engineering & Administrative) i.e.
  - Master Key Switch
  - Beam Path Control
  - Protective Eyewear
  - Posted Signs
- Standard Procedures, review and update
- Training Requirements, do they meet hazards
- Have Exposure Reporting Procedures
- Have Visitor Control Procedures
- FAA and Air Traffic Control Notifications
- Non-Beam Controls

CHECKLIST FOR MANAGEMENT

Ensure the following:

- Relevant safety plans have been approved by internal authority.
- All operator and public safety concerns have been addressed.
- All local authorities that have specific interest and jurisdiction have been notified in case of public interest or concern or possible accident.
- The FAA paperwork is signed off and available if necessary.
- Public perception is not adversely affected by outdoor operations.

Maintain cost and schedule and look for opportunities to:

- Enhance public perception through demonstrations.
- Improve communication with local and federal authority.
- Generate informational and promotional material.

Contact these people:

- Local law enforcement with jurisdiction over test range and flight path
- Local Sherriff's and Police departments
- Local, state, and federal agencies charged with oversight of particular areas
- Entities in charge of parks and forests
- Federal Aviation Administration branch for the region (Western or Eastern)
- Local air traffic control with jurisdiction over flight paths
- Local air traffic control who may wish to issue NOTAMs

16 Laser Light Show Accidents

*Ken Barat**

CONTENTS

The laser light show, since its conception, has been tied to entertainer stage shows.[†] Early shows presented a significant potential for general population injury and yielded a response from regulatory agencies in the United States. Other countries have also responded with an array of national regulations, but for this chapter, we will concentrate on the United States response.

The laser light shows laser sources consist of Class 3B and Class 4 lasers. At one time, these were commonly water or air-cooled Argon and Krypton lasers. Today, those systems and all their associated problems have been replaced by laser diode systems under computer control. Alignment activities are still just as critical, but much of the potential risk has been eliminated. The greatest hazard is still audience scanning.

Early laser art shows evolved around a series of inventions and patents such as what was called Co-Op-Art by Leo Beiser in the early 1960s. Lasers were used in performance art for a production of "Faust" in Stockholm in 1968 and for a ballet at the Opera Comique in Paris around the same time. Photographs of these

---

* With thanks to Roberta McHatton

[†] Laser light shows are not limited to entertainment venues, but are often done in hotel ballrooms, with mirrored walls and chandeliers. These shows are many times part of a business meeting, adding to them a little extra pop!

complex and intriguing patterns had artistic value which were put on display and sold to collectors. One of the first exhibitions of optical transforms was presented by Canadian photographer Lawrence Weissmann at the International Museum of Photography, New York, in 1971. Mr. Weissmann's work differs from most other creators of optical transforms in that he uses images of people and objects rather than geometric shapes.

Laser light took on new dimensions with the magical look created by smoke, mirrors, oscillators and, eventually the addition of X/Y scanners moving beams in the air and on surfaces offering light sculptures set to music. Laser shows became a main attraction, old and young alike; bringing in much needed revenue to planetariums and science centers worldwide.

While lasers were used for rock shows back in the 1980s and 1990s for bands such as Blue Oyster Cult and Pink Floyd, lasers back then were large in size, involving gas tubes or fragile crystals and rods, large/heavy power supplies/exciters needing high voltage power, and water for cooling. On top of these challenges, the lasers themselves were finicky (difficult to fire up and/or went out of alignment easily) thus stressful to work with if not downright dangerous. Not only is laser light dangerous in and of itself but the use of high power required by lasers back then was a constant concern to technicians. While laser beams offered a very dramatic production value—they are the only special effect that truly embraces the audience—the challenge of touring with them saw a decline in use as other lighting technologies such as LEDs and Video Projection became available. That all changed with the advent of Diode, OPS, and OPSL lasers rendering projectors capable of emitting 25 W—downsized such that it is about the size of a breadbox—which can be plugged into a wall socket and offered more colors better balanced than gas lasers. New improved software also made the most of the new laser technology.

LASER ACCIDENTS

Laser users need to be aware that laser accidents do happen, many times for reasons that could have been prevented. Laser accidents at light shows could involve the operator or audience. If the researcher in a lab has an incident or accident, it is rare that more than one person is injured. While an incident at a light show has the potential to injure a vast number of people, even thousands (which luckily has not happened yet). It is this potential that has driven regulatory action at the Federal and State level. But accidents have occurred.

There are regulations and guidelines for setting up laser light shows with the goal of making a safe and enjoyable event. But things have gone wrong. Here are a few examples for your review:

CASE 1: LAS VEGAS AIRLINER STRUCK WITH LIGHT SHOW BEAM

Report #285091 stated: "[On takeoff], at approximately 500 feet AGL, a laser beam of green light struck through the right-side window of my cockpit striking my First Officer in the right eye and blinding both he and I for approximately 510 seconds due to the intensity of the light beam. I immediately notified the Tower Controller [who stated] that this had become a recurring problem with the laser show coming from the top of the [hotel] in Las Vegas. We were very fortunate, because this could have been a much more serious situation had the laser struck myself as well as [my First Officer] at a more direct angle, severely blinding both of us and endangering the lives of my passengers and crew."

CASE 2: MOSCOW RAVE INCIDENT 2008

Dozens of partygoers at an outdoor rave near Moscow lost partial vision after a laser light show burned their retinas. Moscow city health department officials confirmed 12 cases of laser-blindness at the Central Ophthalmological Clinic, and daily newspaper *Kommersant* said another 17 were registered at City Hospital 32 in the center of the capital. Attendees at the July 5 Aquamarine Open Air Festival in Kirzhach, 80 km (50 miles) northeast of Moscow, began seeking medical help days after the show, complaining of eye and vision problems, health officials told *Reuters*. "They all have retinal burns, scarring is visible on them. Loss of vision in individual cases is as high as 80 percent, and regaining it is already impossible," *Kommersant* quoted a treating ophthalmologist as saying. Attendees said heavy rains forced organizers to erect massive tents for the all-night dance party, and lasers that normally illuminate upward into the sky were instead partially refracted into the ravers' eyes. "I immediately had a spot like when you stare into the sun," rave-attendee Dmitry told *Kommersant*. "After three days, I decided to go to the hospital. They examined me, asked if I had been at Open Air, and then put me straight in the hospital. I didn't even get to go home and get my stuff," he said. The owner of a Moscow laser rental company told *Reuters* that the accidental blindings were due to "illiteracy on the part of technicians."

CASE 3: MYLAR BALLOON INCIDENT

A laser light show operator was struck by a reflected beam from a Mylar balloon let loose by an audience member during an outdoor show. From the operator's perspective, luckily no audience member was hit. Minor eye injury was sustained.

CASE 4: DANCER GRAZED BY LASER CONE EFFECT

The performer was standing on a pressure mat. The mat was set to terminate beams once the performer stepped off. A cone of beams surrounded the performer. When he stepped back off, the beams did not go off. Because there was an observer, the beams were manually terminated. What happened? The mat was just where it was supposed to be, but no one checked to see if it was working. It was not an equipment failure; it was just not plugged in to the laser system.

CASE 5: ARGON IN THE EYE—1990

An argon laser was set-up on a tower and was designed to strike the side of a building. It was part of a promotional plan to bring people to downtown Phoenix. A mask was set-up to prevent it from aiming too low or high. During the nightly alignment, the operator received a 20 W argon reflection in the eye. The operator, upon arriving at the tower, realized he had left his laser protective eyewear at home. Having performed this alignment before, and with a "show must go on" attitude he attempted the alignment without the use of eyewear, yielding him no protection when the reflection occurred. Significant eye damage occurred.

CASE 6: FIRE INSPECTOR INJURED

A fire inspector in a city that requires a special laser permit was conducting an inspection without adequate safety training. He did not confirm stability of a laser beam block. While standing against a wall, a laser was terminating onto it to ascertain if there were reflective surfaces. The beam block slipped and he was lasered in his cornea resulting in a fairly severe burn that took a couple of months to heal. Fortunately, the laser did not hit the lens of his eye! Lesson learned was he should have been provided with laser safety training that included checking beam blocks were tightened before standing in front of the laser.

CASE 7: POOR PLACEMENT

All lasers were programed for very specific placement of a thrust stage. When the stage was set-up during a live performance, it was one foot off the mark. The laserist was not informed, thus unaware until lasers were hitting the heads of about 19 people in the front row and the performer and was not able to shutter the lasers in time to prevent unwanted exposure. Thankfully, no one looked up; the lasers were only 80 mW and no one was harmed.

CASE 8: TOMORROWLAND FESTIVAL IN BELGIUM–FIGHTING PERCEPTION AND HEADLINES

Two people claimed injury at a music and dance festival in Boom, Belgium. Investigation findings indicate they were not injured by lasers from the light show but rather from handheld lasers. In the news report, one individual was reported to have received irreversible injury to the central portion of their eye. This type of statement could be related to the fact that most doctors have a lack of experience with laser eye injuries. Calculations of the beam effects make this diagnosis unlikely.

A police investigation found there were a number of laser pointer-like devices being used by audience members. Some were found to have an output of 200 mW, considerably over the 5 mW laser pointer level. At least five individuals were observed with what we should term handheld lasers. The MPE for a visible continuous wave laser is 2.55 mW/cm^2. It was stated that the irradiance for the handheld lasers used by some audience members might have been 1020 mW/cm^2 or greater.

One item to remember is many more will remember the line "irreversible vision loss," read the follow-up story, and blame some individuals in the crowd misusing handheld lasers.

CASE 9: WHITNEY ART MUSEUM "LIGHT SHOW"

A "light show" that reportedly caused injury to spectators at the Whitney Museum in December 1980 was the subject of two lawsuits that had lawyers searching for precedents. One lawsuit charged that a show of light works by the artist James Turrell created an illusion, whereby a woman became disoriented and confused and was "violently precipitated to the floor" of the museum, according to the legal papers, thereby sustaining "severe and permanent personal injuries." The woman suffered a broken arm. The suit also sought damages in an unspecified amount from the artist.

The other suit charged that the same exhibition caused another woman, after stepping back against what she thought was a wall, to fall and permanently injure her right wrist. She asked for damages of $250,000 from the Whitney Museum.

The show, titled "James Turrell: Light and Space," exhibited the work of the well-known West Coast artist, who is concerned with the effects of light in space. Mr. Turrell describes his work as "making light inhabit space so it feels materially present," and the show presented such works as a solid wall that seemingly held a gray painting; in reality, a rectangular cut through the wall filled with gray light. Other installations gave the illusion of solid cubes or shapes that, when approached closely, revealed themselves as composed entirely of light.

Mr. Turrell said he felt blameless. "I'm accused of creating what was created," he said. "It was a very quiet show; the work isn't hazardous. The intention is to change one's thinking about seeing. I'm not responsible for how someone else takes care of his or her sense of bodily awareness."

The Whitney Museum officials said they had no comment on the suits, other than that the matter was in the hands of the museum's insurance company. Several New York lawyers said they could not remember a precedent for such a suit. "If a case of this sort were allowed to succeed, any artist who dealt with art work of a powerful nature likely to cause a strong reaction in the viewer would have to worry about the depth of that reaction. It seems to me to be an impossible restraint to place on an artist."*

AUDIENCE SCANNING

As stated, regulatory rules exist for laser light shows. If they are followed, the likelihood of a laser accident is extremely rare. There are requirements for beam separation from the audience. However, all that changes during audience scanning. Regulatory agencies take audience scanning very seriously and additional safe guards are required. The audience is either scanned directly from the laser source or a laser being reflected off a rotating target. The most common of these is a famous disco mirror ball.

The following is an explanation of why mirror balls work and is based on CDRH guidance documents (Figure 16.1).

---

* Grace Glueck, "Whitney Museum Sued over 1980 'Light Show,'" *New York Times*, May 4, 1982.

FIGURE 16.1 (a) Mirror Ball, example 1, (b) Mirror Ball, example 2, (c) Audience Scanning, example 1, and (d) Audience Scanning, example 2.

MIRROR BALLS

Mirror balls are frequently used in light shows to separate and reflect the laser beam into many rays of laser light. When done properly, this can significantly reduce the power and, therefore, the potential hazard of a laser beam. If the beam is reflected off enough facets on the mirror ball, the resulting rays will go off in many directions. Although the individual rays still do not diverge very much, each has only a fraction of the power in the direct beam. Obviously, the degree of safety that this can produce depends upon the power of the direct laser beam, and the number of rays and directions into which the beam is split.

The more rays into which the beam is split, the smaller the fraction of power each reflected ray will have. A scanning device is usually used to sweep the beam back and forth across a broad section of the mirror ball so that the beam is broken up by several facets on the ball. Rotating the mirror ball can provide even more safety because the movement of the reflected rays reduces any exposure time. Without a scanning device, or without a properly designed scanning system, the beam is broken by the mirror ball into fewer rays, each having a larger fraction of the power in the direct beam. This means that even with a mirror ball, there could still be a potential for harm.

AUDIENCE SAFETY MEASUREMENTS

For audience safety measurements, you need to accurately measure the beam's irradiance. This is the power over a given area, expressed in units such as "milliwatts per square centimeter."

For example, a 1-watt laser can put all of its power into a single thin beam that can enter the pupil. The irradiance is very high; it can instantly cause severe eye damage. Or, the same beam can be spread out using lenses or scanning, so that only part of the light enters the eye at any one time. This lowers the irradiance. If the irradiance is low enough, the beam from a 1-watt laser can be safe for audience exposure. Irradiance measurement need calibrated instruments; making a visual judgment will only lead to injuring to someone.

17 Fiber Optics in Telecommunications

Larry Johnson

CONTENTS

Today, fiber-optic communication systems (FOCS) allow for voice, video, and data communications to occur at the speed of light and it is the laser that provides the efficient light source that drives this industry.

The standard wavelengths used in fiber-optic communication systems are the 850 nanometer (nm) window, 1300 nm window, and the 1550 nm window. The term window is used loosely in the fiber-optics industry but ranges from plus or minus 30 nanometers from a specified wavelength, so a 1300 nm system could have a laser transmitting anywhere with a center wavelength from 1270 to 1330 nm. It should also be noted that all these wavelengths are infrared, which presents safety issues for certain applications and equipment.*

FIBER-OPTIC SAFETY STANDARDS

The evolution of fiber-optic components, systems, and technology to address data rates in the gigabits requires the use of laser diodes in most applications due to speed requirements. Even short distance multimode systems, which in the past used LEDs, have now migrated to low cost Vertical Cavity Surface Emitting Laser sources (VCSEL). For this reason, all fiber systems must be treated as if they are carrying a laser signal. Attention on how to address Laser safety and the safe design, use, and implementation of lasers is required to minimize the risk of eye accidents. Laser classifications are based on the concept of accessible emission limits, or AEL. This is usually a maximum power level in watts, or energy level in joules that can be emitted at a specific wavelength and exposure time.

In the United States, lasers are regulated by the Center for Devices and Radiological Health, or CDRH. A branch of the FDA. the CDRH is responsible for overseeing the

---

* Many times, the LSO fails to realize that his own institution may have staff who are installing and maintaining its fiber-optic communication network. Many of these have moved from LED light sources to laser diode sources, some of which are emitting in the Class 3B range. Therefore, there is an obligation to look at their safety needs and practices (Ken Barat).

manufacturing, importation, performance, and safety of all medical devices as well as devices that emit certain types of electromagnetic radiation, including lasers. With lasers, the CDRH is concerned that the devices are properly labeled as to their output power and are equipped with the appropriate safety equipment if necessary.

A number of organizations have developed standards and guidelines for safely working with optical fiber, cables, and optical transmission equipment. These include the ANSI Z136.2 American National Standard for the *Safe Use of Optical Fiber Communication Systems Utilizing Laser Diode and LED Sources* and also the OSHA standard on laser safety STD-01-05-001.

For international use, the IEC 60825-1 covers the safety of laser products, and the IEC 60825-2 covers the safety of optical fiber communications systems (OFCS).

The ANSI Z136.2 and the IEC 60825-2 standards divide laser devices into a set of four general classes and several subclasses based on their wavelength and optical power output. The first version of the standard was published in 1988 at a time when laser-based fiber-optic communications systems were very simple. They used only two wavelengths, 1310 and 1550 nanometers, at maximum power levels well under 10 milliwatts. The standard was the first written specifically for fiber-optic communications systems.

Under the ANZI 136.2 standard, an end-to-end fiber-optic system is considered a Class 1 laser product because under normal conditions the laser emissions are completely enclosed. It is not until the fiber is broken or a connector is unplugged that a person may be exposed to laser radiation which may be potentially hazardous. Therefore, the hazard level for each optical port must be individually assessed, and control measures defined. To address such situations, the standard defines the hazard level as the potential optical hazard at any accessible location within a fiber system, depending on the level of the radiant energy that could become accessible. Hazard levels are assigned values from 1 to 4 based on the customary laser accessible emission limits, or AELs.

For example, Hazard Level 1 does not exceed the AEL for laser Class 1, and so forth. If automatic power reduction (APR) is used, the normal level of power in the fiber and the speed of the APR system determine the hazard level.

A Class 1 laser system is considered to be incapable of producing damaging radiation levels during operation, and is exempt from any control measures or other forms of surveillance. Class 1 laser products are incapable of producing damaging radiation levels to eyes or skin during operation, including direct viewing of the laser beam with optics that could concentrate the laser output into the eye. Thus, power levels must be less than 4 microwatts. Class 1 systems are exempt from labeling or any other control measures and surveillance. However, if a Class 1 system containing an enclosed laser of a higher power level is opened, then interlocks are required to shut down or reduce the laser power to Class 1 levels. All LED light sources and many VCSELs used with multimode fibers are Class 1 light sources.

A Class 1M laser is incapable of producing hazardous exposure conditions during normal operation unless the beam is directly viewed through an optical instrument such as a loupe or a fiber inspection scope. For this reason, the "M" stands for "magnifying optics caution." Class 1M lasers are also exempt from control measures other than to prevent direct viewing of the beam through optical instruments.

Many fiber-optic communications lasers are Class 1M including VCSELs, Fabry-Perot (F-P), and distributed feedback (DFB) types operating at 850, 1300, 1310, and 1550 nanometers. Class 1M lasers are also exempt from control measures other than to prevent direct viewing of the beam through optical instruments.

Class 1M lasers produce large-diameter beams, or beams that are divergent, such as the beam emitted from the end of an optical fiber. 1M is a new class that applies to a wide wavelength range from 302.5 to 4000 nanometers, and having a total output power below the Class 3B level, but the available power that can pass through the pupil of the eye must be within Class 1. Most FOC systems do not require optical amplification and are typically low levels and for this reason most FOC systems use low power Class 1 light sources.

Class 2 lasers emit in the visible part of the spectrum between 400 and 700 nanometers where eye protection is generally provided by a person's natural aversion response to bright lights. Optical power levels can be up to 1 milliwatt, so many laser pointers are included in this class.

Class 2M is a newer class that include lasers emitting in the visible part of the spectrum, and also rely on aversion responses for eye protection, but can be hazardous if viewed through optical instruments. Laser pointers can also be in this class as well as many fiber identifiers. The 5-milliwatt red laser diode fiber tracers are Class 2M.

Class 3, or medium power, lasers are divided to two sub-classes: 3R and 3B. Class 3R lasers have reduced product safety requirements and represent a transitional zone between safe and hazardous laser products. Optical power levels can range from 1 to 5 milliwatts. Although the risk of eye injury from direct viewing remains relatively low, greater efforts should be taken in the use of these lasers.

While the "B" has no special meaning other than a means of dividing Class 3 lasers, the "R" stands for Reduced Requirements. Class 3R lasers have reduced product safety requirements and represent a transitional zone between safe and hazardous laser products. A Class 3 laser may be hazardous when directly viewed, and its specular reflection off a shiny surface may also be hazardous. In fiber optics, most optical amplifiers are Class 3 lasers. Class 3B lasers include continuous-wave devices with power levels ranging from 5 milliwatts to 500 milliwatts, or energies of up to 125 millijoules for pulsed devices. Class 3B lasers may be hazardous to the eyes for any direct exposure, even with aversion responses. For the higher power levels in this class, the specular reflection from a shiny surface may be hazardous and may also pose a skin hazard, but aversion responses to localized heating generally prevents a skin burn. Diffuse reflections off surfaces like walls or paper are generally not hazardous. Protective eyewear is typically required where direct viewing of a Class 3B laser beam may occur and these lasers must be equipped with a key switch and a safety interlock that shuts the laser down when there is a danger of human exposure.

Class 4 covers high power lasers ranging from 500 milliwatts to hundreds of kilowatts for continuous-wave devices, or energies greater than 125 millijoules for pulsed lasers. Class 4 lasers are high power lasers whose output is a hazard to eyes or skin from the direct beam, specular reflections, and in some cases even diffuse reflections. They often have sufficient optical power to be a fire hazard as well.

Engineering and administrative controls are specified for the hazard level and type of environments in which the fiber system may operate. Hazard Level 4, analogous to a Class 4 laser, is not permitted in any location. This means that any fiber system carrying Class 4 optical power levels needs to incorporate control mechanisms to reduce the power to an acceptable hazard level during an event that would permit access to the radiant energy from a fiber or cable.

For each of the hazard or access levels, the type of location will determine the appropriate control measures employed. There are four location types defined in the ANSI standard. An "unrestricted" location is where access to the protective housing, or the open beam of the communications system is unrestricted such as in domestic premises or locations open to the general public.

A "restricted" location is where access to the protective housing or open beam is not accessible to the public. Examples include secured areas in industrial facilities, or overhead cables and fiber-optic cable drops to a building. In the case of Fiber-to-the-Home installations, the upstream laser has a safety interlock which shuts down the upstream laser if the connection is opened.

A "controlled" location is where access to the protective housing is controlled and accessible only to authorized persons who have received adequate training laser safety and servicing of the system. Manufacturing and laboratory sites are common examples of controlled locations.

When the hazard or access class as well as the location type is known, the appropriate safety control measures can be applied. These measures include the correct labeling of optical ports with highly visible tags, sleeves, tape, or signs containing a warning message based on the hazard level of the potential exposure to laser energy if the level exceeds Class 1.

When operating lasers of Classes 3B and 4, in a manner that may result in eye exposure in excess of the maximum permissible exposure limit, protective eyewear is required in the workplace by the U.S. Occupational Safety and Health Administration.

The safety hazard posed by any laser depends on a combination of wavelength, intensity, beam divergence, and exposure time. This concept is the key to the laser safety standards implemented around the world.

These advancements have allowed modern optical communication systems to pose hazards significantly greater than in the past. This is why standards must continually evolve to meet current needs. In 2007, the IEC 60825-1 standard featured a reorganization of the laser classification system, where a number of subclasses have been added to better define the hazards posed by laser devices. This was further updated using the optical fiber communication systems as defined in the IEC 60825-2 standard issued in 2010 and the ANSI 136.2 standard in 2013.

OPTICAL SAFETY IN FIBER-OPTIC SYSTEMS

When properly installed and operating, most fiber-optic communication systems are Class 1 laser systems because laser energy is low power and is completely enclosed or contained during normal operation. However, when these systems are serviced or repaired, it is possible that workers may be exposed to hazardous levels of laser radiation. This is especially true of systems employing Raman amplification or DWDM networks with large channel counts.

The ITU-T G.664 *Optical Safety Procedures and Requirements for Optical Transmission Systems* recommendation defines a set of optical safety procedures and requirements primarily aimed at optical transport networks. The recommendation focuses on Automatic Power Reduction (APR) techniques with automatic restart.

APR techniques are employed to reduce laser power to a safe level when there is a loss of optical continuity in the main optical path. This could be caused by a cable break, equipment failure, an unplugged connector, or similar occurrence. It is also recommended to use APR with automatic restart, and not with manual restart methods. In the past, APR used periodic full power pulses to test for link continuity, but this is no longer considered appropriate.

One current approach uses an Optical Supervisory Channel with a normal operating power at Hazard Level 1 or 1M, so it can be kept alive on the fiber after the main laser power has been reduced. Restoration of continuity on the supervisory channel is then used to bring back full operational power to the fiber. In this way, it is ensured that full power is only present in a fully enclosed configuration, thus guaranteeing optical safety.

Multi-fiber array connectors such as the MPO type can handle up to 24 single-mode fibers or up to 72 multimode fibers in a single termination. These high fiber counts along with number of potential wavelengths that may be optically multiplexed on each fiber can greatly increase possible safety hazards. With the N.A. of a fiber, where light exits the optical fiber, there is a high chance that the eye could "see" not only multiple fibers but multiple wavelengths over each fiber when looking at or inspecting array connectors.

When fibers are carrying high optical power levels, the connectors or even the fibers themselves could be damaged under certain conditions. These include loss-induced heating at connectors and splices, as well as connector end-face damage caused by burning of dust or other contaminants on the optical surfaces.

Tight bends in fibers carrying high power levels can cause localized heating, result in the melting or burning of the fiber coatings. In severe cases, these localized heating effects could lead to fires, so safety practices for high power systems must not be taken lightly.

The ITU-T G.39 *Optical System Design and Engineering Considerations* recommendation describes the best practices for optical power safety. Among the topics covered, are the proper termination, cleaning, and viewing of potentially live optical fibers.

At high power levels, proper cleaning of optical connectors is of paramount importance to avoid potentially hazardous heating effects and possible damage. Use only cleaning methods approved by the operating organization and only clean connectors when it has been verified that the fibers are not in use (these are often called "dark" fibers).

Technicians must always treat unknown fibers or optical ports as if they are carrying optical signals. In addition, optic cables should be treated as if they are carrying laser signals unless it has been verified that the optical transmitter is turned off or that all fibers in the cable are dark or at a safe optical power level.

A number of methods can be used to determine the operational status of a fiber or optical port. One of the simplest methods is the use of an optical power meter. However, using a power meter on a broken fiber may be difficult because they are usually designed to work with fibers that have been terminated by connectors. A bare fiber adapter can be used to adapt the fiber to the connector type used on the power meter.

One potential problem with the power meter can be its sensitivity. If the fiber is carrying high-power laser energy, it may oversaturate the power meter's detector. This could result in the meter displaying its maximum rated power reading, when the fiber is actually carrying significantly more optical power. From this false information, the technician could conclude that the fiber is emitting safe levels of optical energy when in fact a safety hazard exists. High optical power levels can also damage a power meter unless it has been designed for high power use, so care must be taken when using low-power meters on unknown fibers.

Power meters must also be set to the correct wavelength for accurate readings. If the technician does not know what wavelengths are in use, erroneous readings may lead to a safety hazard. Newer power meters as well as those designed specifically for DWDM use may have automatic wavelength selection.

It is up to equipment manufacturers to implement safety safeguards in the products they manufacture. However, users should verify what safeguards are present and how to correctly configure them for use in specific network safety applications.*

DETECTING LIVE FIBERS

A simple method to quickly detect live fibers is the use of an infrared viewer (Figure 17.1). The viewer is a handheld instrument that uses a special type of electron tube that collects infrared light from a lens at one end of the device, converts it to a stream of electrons, and projects an image onto a phosphor viewing screen at the other end. By projecting the light from a fiber or cable onto a scrap of paper, the technician can see the average brightness of the light as well as its divergence as the fiber or cable is moved closer to and further away from the paper.

Video-based fiber inspection scopes can also be used to detect live fibers. They provide maximum eye safety as laser light from a fiber strikes the video imager rather than the technician's eye.

Another simple and inexpensive method of detecting live fibers is the use of an infrared viewing card (Figure 17.2). The cards consist of phosphors that are sensitive to infrared radiation coated on a substrate and laminated with clear plastic. After "charging" the phosphor coating in sunlight or room light, the card is positioned near the end of the fiber to be tested. The presence of infrared radiation will cause the phosphor to glow, indicating that the fiber is live. The cards are inexpensive, passive devices that can easily be carried in a shirt pocket and can provide a simple, go/no-go test for live fibers.

---

* See Chapter 19 for a review of non-beam hazards fiber optics present to the users.

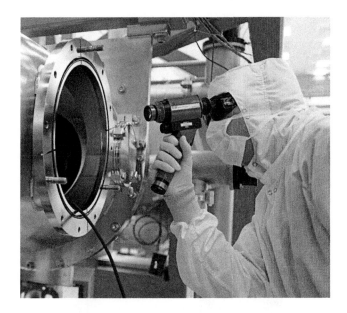

FIGURE 17.1 Infrared Radiation viewer.

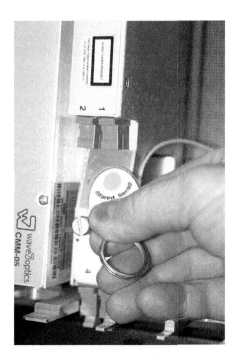

FIGURE 17.2 Infrared Radiation sensor card. (Courtesy of The Light Brigade.)

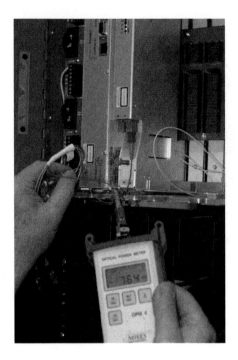

FIGURE 17.3   Output meter, example 1.

Simple detection methods, however, will not reveal the amount of optical power carried by the fiber. For this, an optical power meter is still the best choice (Figures 17.3 and 17.4).

Whereas the construction hazards of a fiber-optic transmission system are temporary, the potential optical hazards will exist throughout the entire working lifetime of the system, and will affect not only the installation personnel, but future maintenance technicians and even end-users as well.

A broken fiber will usually cause laser energy to be more divergent, but cleaving live fibers can result in specular reflections as well as a tighter beam exiting the fiber as determined by the fiber's N.A. For these reasons, the technician must never look directly into the end of an open fiber-optic cable and to determine if the fibers to be spliced or terminated are carrying laser energy. It is best to work on dark fibers, but this may not always be under the technician's control.

SUMMARY

Most fiber-optic communication systems use Class 1 light sources and have low enough optical power not to provide a safety issue. However, most technicians do not know which fibers are transmitting low versus high-power lasers and optical amplifiers. The best practice is to never look into the end of an optical fiber scope until the optical power level has been confirmed.

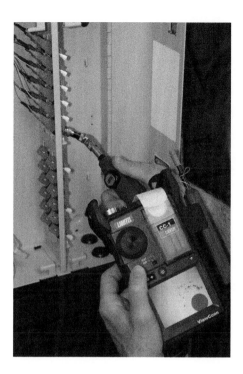

FIGURE 17.4 Output meter, example 2. (Courtesy of The Light Brigade.)

Working with fiber-optic systems and components requires the knowledge of the type of light source used, its wavelength, and optical power level. The safety program may vary depending on whether the user is in a laboratory, manufacturing environment, outside plant, or indoor installation. The safety and system standards are always evolving to meet the safety requirements of the present and future. Even though most fiber-optic communication systems are low powered, new technologies requiring higher optical power levels, amplification, and optical multiplexing are transmitting with Class 3 and Class 4 power levels.

18 Accident Investigation

Ken Barat

CONTENTS

SETTING THE SCENE

A laser incident has occurred, for real or suspected. You have been notified, and the person involved has been dealt with (medical examination). Now it is time to find out what happened.

This will either fall to the LSO or in some institutions, a group of individuals trained in accident investigation. For most, the latter is a luxury they do not have, so once again, it falls back to the LSO.

To be prepared for such an event, the LSO should have already put together a pool of experts, laser users, an electrical safety officer, interlock specialist if one is present, and so on. The goal of this pool is to have at the LSO's disposal resources to assist in any accident investigation.

There are many approaches to accident investigation as well as training courses on each technique. Having been involved in several approaches, I really do not know which is the best approach. But there are a few items that are essential for the LSO to be aware of:

1. Do not go into any investigation thinking you already know all the facts. This approach can blind you to what really happened as well as biases your approach.
2. Try not to ask questions that will yield one-word answers such as Yes, No, Correct, and so on.
3. Plan on taking pictures.
4. Ask for written statements by the individual and those around the individual
5. Beware that at times there will be a great deal of pressure to allow operation to continue as soon as possible or even during your investigation. Decide ahead of time what your position will be on this.

The goal of the investigation is to find out what caused the accident and determine what can be done to prevent it from happening again. It is not to find out whose head should roll.

The hardest accident investigation to investigate and get the facts out of is the single person event: just one person and no one around. This is why having a laser user to ask about how common the person's actions were and what is a good technique is so useful. But the "pilot error" setting is the most common in the research lab.

PREPARATION

Just like many first responders have a response bag set aside ready to go, the LSO needs to be ready. Your ready kit need not be overwhelming—power meters are not required—but here are a few critical items:

- List of people to call on for assistance, users and medical staff, phone numbers and extensions
- Digital camera
- Signs indicating no entry or changes to set-up allowed
- Tape measure
- Writing or electronic tablet
- Optional items: Senor cards, IR viewer

Everyone these days has a camera in their phone, and while this is adequate to take images at the scene, a digital camera might be better. Chiefly due to the option of using wide angle lens, and greater zoom capability. Here are some hints on taking images:

- Always make notes about the photos taken
- Start by taking distance shots first then move in to take closer photos of the scene
- Take photos at different angles (from above, 360 degrees of scene, left, right, rear) to show the relationship of objects and minute and/or transient details such as ends of broken rope, defective tools, drugs, wet areas, or containers
- Take panoramic photos to help present the entire scene, top to bottom, side to side
- Take notes on each photo; these should be included in the incident investigation file with the photos
- Identify and document the photo type, date/time/location taken, subject, weather conditions, measurements, and so on
- Place an item of known dimensions in the photo to add a frame of reference and scale (e.g., a penny, a pack of cards)
- Identify the person taking the photo
- Indicate the locations where photos were taken on sketches

The Occupational Safety and Health Agency (OSHA) has a free accident investigation guide; the following is from that guide.

OSHA strongly encourages employers to investigate all workplace incidents—both those that cause harm and the "close calls" that could have caused harm under slightly different circumstances. Investigations are incident-prevention tools and should be an integral part of an occupational safety and health management program in a workplace. Such a program is a structured way to identify and control the hazards in a workplace, and should emphasize continual improvement in health and safety performance. When done correctly, an effective incident investigation uncovers the root causes of the incident or "close call" that were the underlying factors. Most important, investigations can prevent future incidents if appropriate actions are taken to correct the root causes discovered by the investigation.

Effective incident investigations are the right thing to do, not only because they help employers prevent future incidents, but because they help employers to identify hazards in their workplaces and shortcomings in their safety and health management programs. Investigations also save employers money, because incidents are far more costly than most people realize. The National Safety Council estimates that, on the average, preventing a workplace injury can save $39,000, and preventing a fatality more than $1.4 million, not to mention the suffering of the workers and their families.

The more obvious financial costs are those related to workers' compensation claims, but these are only the direct Incident Investigations: A Guide for Employers, December 2015. "One central principle ... is the need to consider the organizational factors that create the preconditions for errors as well as the immediate causes."* The indirect costs are less obvious, but very commonly greater, and include lost production, schedule delays, increased administrative time (for emergency response, investigations, claim processing, and others), lower morale, training of new or temporary personnel, increased absenteeism, and damaged customer relations and corporate reputation.

The National Safety Council is another excellent source of guidance and material on accident investigation: Here are 9 steps of the investigation process:

1. Call or gather the necessary person(s) to conduct the investigation and obtain the investigation kit.
2. Secure the area where the injury occurred and preserve the work area as it is.
3. Identify and gather witnesses to the injury event.
4. Interview the involved worker.
5. Interview all witnesses.
6. Document the scene of the injury through photos or videos.
7. Complete the investigation report, including determination of what caused the incident and what corrective actions will prevent recurrences.
8. Use results to improve the injury and illness prevention program to better identify and control hazards before they result in incidents.
9. Ensure follow-up on completion of corrective actions.

---

* Sidney Dekker (2006) costs of incidents.

WHAT IS THE PURPOSE OF THE ACCIDENT INVESTIGATION?

This is an important question to ask. We have already stated it is not to find blame but rather what happened and the related causes and how they can be corrected. Often termed causal factors.

Factors that might have contributed to the event may have involved equipment, the environment, people, and management. There are a number of well-established causal factor questions that apply to the majority of accident scenarios, which can be applied to laser incidents:

1. Did a written or well-established procedure exist for employees to follow?
2. Did job procedures or standards properly identify the potential hazards of job performance?
3. Were there any non-beam or hazardous environmental conditions that may have contributed to the incident?
4. Were any actions taken by employees, supervisors, or both to eliminate or control environmental hazards?
5. Were employees trained to deal with any hazardous environmental conditions that could arise?
6. Was sufficient space provided to accomplish the job task?
7. Was there adequate lighting to properly perform all the assigned tasks associated with the job?
8. Were employees' familiar with job procedures?
9. Was there any deviation from the established job procedures?
10. Were the proper equipment and tools available and being used for the job?
11. Did any mental or physical conditions prevent the employee(s) from properly performing their jobs?
12. Was there anything different or unusual from normal operations? (e.g., different parts, new or different chemicals used, recent adjustments/maintenance/cleaning on equipment)
13. Was the proper personal protective equipment specified for the job or task?
14. Were employees trained in the proper use of any personal protective equipment?
15. Did the employees use the prescribed personal protective equipment?
16. Was personal protective equipment damaged or not properly functioning?
17. Did supervisors and employees participate in job review sessions, especially for those jobs performed on an infrequent basis?
18. Were supervisors made aware of their responsibilities for the safety of their work areas and employees?
19. Were supervisors properly trained in the principles of incident prevention?
20. Was there any history of personnel problems or any conflicts with or between supervisors and employees or between employees themselves?
21. Did supervisors conduct regular safety meetings with their employees?
22. Were the topics discussed and actions taken during the safety meetings recorded in the minutes? Were the proper resources (i.e., equipment, tools,

materials, and so on.) required to perform the job or task readily available and in proper condition?

23. Did supervisors ensure employees were trained and proficient before assigning them to their jobs?

THE ART OF THE INTERVIEW

You may wind up only interviewing the person injured (working by themselves). There may be others who can contribute to understanding why things happened. A person you may not think of is the person who trained the injured person. This may tell you if the individual injured was properly prepared for the activity they were engaged in. Thinking about the note at the start of this chapter, we want to have people in a relaxed non-threatening position. Learn to reduce their understandable anxiety. Here are some established guidelines for the art of the interview:

1. Conduct the interview in a quiet and private place
2. Use open-ended questions
3. State that the purpose of the investigation and interview is to find out what happened, not who is at fault
4. Ask the interviewee to recount their version of what happened without interrupting. Take notes or record their response
5. Ask clarifying questions to fill in missing information
6. Don't be afraid to ask what might seem as naïve questions
7. Repeat back to the interviewee the information obtained. Correct any inconsistencies
8. Ask if they think could have prevented the incident, focusing on the conditions and events preceding the injury
9. Thank the person for their time and cooperation
10. Finish documenting the interview

DOCUMENT YOUR FINDINGS

Remember, the manager who reads the report does not want a novel. They want to know what happened, why, and how to correct it. Everything included in the report should be supported by facts and the evidence you collected. Start with an executive summary or short narrative. This should be followed by a detailed and more complete account of the accident.

CORRECTIVE ACTIONS

The old statement "The Facts, Just the Facts" is not sufficient in an accident report. The most important thing is the action needed to prevent a reoccurrence. For each corrective action, it should be clear who is responsible to carry them out.

LESSON LEARNED

The accident report, no matter how well written, fails if a lesson learned is not distributed to the appropriate staff.

FOLLOW-UP

Closely related to "Corrective Actions" is the follow-up. The LSO needs to follow-up on corrective actions to see that they are occurring (Table 18.1). This can be a real challenge. Especially when corrective actions cost money. The question always becomes: Whose money?

TABLE 18.1
Checklist for Incident Investigations/Reviews

Incident Investigation Elements	Checklist Items	Considerations
1. Plan the Review	Identify the basic information related to the incident	For injury incidents, review any initial or medial reports
	Select the Review Team	Team composition will vary based on the category of the injury and the complexity of the investigation. • Supervisor (typically leads investigation) • Employee • Division Safety Coordinator • EH&S Liaison • EH&S Incident Investigator • Subject Matter Expert-LSO
	Preserve the incident site	Do not disturb the incident site while the review is in progress
	Schedule the interviews and visit to the incident site that optimizes time usage.	Initial report should be completed within 7 days
2. Collect Data	Observe the incident site and record the conditions	a. Do not alter or modify the incident scene. b. Note if the scene has been altered c. Examine the incident scene: equipment, work space, engineered controls, PPE, and so on, and take notes. d. Obtain photographs of the incident scene and when safe, photograph an incident re-enactment e. Review any written procedures or documents related to the incident. f. Maintain an open mind as you collect the data.

(Continued)

TABLE 18.1 (*Continued*)
Checklist for Incident Investigations/Reviews

Incident Investigation Elements	Checklist Items	Considerations
	Interview Process	Start the interview by explaining the reason for the interview and focus on the positive objective of learning from the incident to prevent recurrence.
		Ask for a description of what happened from start to finish.
		Ask questions for clarification.
		Start at end of the sequence and work backwards looking for gaps and asking for other information that has not yet been shared.
		Close with a thank you and provide information as to what happens next in the investigation process.
	Tips for conducting interviews	• Apply "caring objectivity"
		• Avoid asking why, and instead ask how and what
		• Do not interrupt!
		• Try to stimulate recall
		• Be curious and inquisitive but suspend judgment
		• Use open-ended questions
		• Do not attribute blame
		• Enlist the help of the interviewee to provide information that will prevent recurrence of the incident.
		In a group, only one interviewer at a time should ask questions
	Some typical interview questions	• What was the sequence of events and conditions leading up to and during the incident?
		• Was there anything abnormal or any recent changes?
		• Were there any workarounds and/or task compensations?
		• Were there previous related occurrences?
		• Are there procedures and how are they related to the incident?
		• What is the training status for the activity?
		• How much experience does the organization and individual have with performing the activity in question?
		• What were any contributing factors? How could it have been prevented?

(*Continued*)

TABLE 18.1 (*Continued*)
Checklist for Incident Investigations/Reviews

Incident Investigation Elements	Checklist Items	Considerations
3. Assemble and Review the Data	Define the incident to be reviewed.	• The incident is the reason for the investigation. • For first aids and OSHA recordable cases, the incident typically is the injury.
	Group data into Events and Conditions.	• Events: "Who did what?" or "What did what?" (active verbs) • Each event should include only one action • Conditions (What we know about the event.); • Events and conditions should be based on facts and not assumptions.
	Establish the order of events and conditions leading up to and during the incident	• Construct a diagram showing the chronological order of events up to the incident and then add conditions to the events. • Make a time-ordered chart • Diagram should include dates and times • Use job titles instead of staff names
	Review time order of events for the incident.	• Have you identified the sequence of events • What happened (including when and where the incident took place, and who was involved) • Have you identified the conditions related to each event? (What we know about the event? What were the activities, events, objects, substances, and damages?).
	Review the diagram for completeness and reasonableness.	• If important data gaps exist, collect more information.
4. Identify Root Causes and Corrective Actions	Review results from Step 3 and apply appropriate root cause analysis techniques—Key terms	• Causal Factor: Any problem or issue that, if corrected would have prevented the incident from occurring or significantly reduced the incident's consequences. • Root cause: The most basic cause of causes that can reasonably be identified that management has control to fix and when fixed, will prevent (or significantly reduce the likelihood of) the problems recurrence. • Corrective Action: A change (procedure, policy, equipment, and so on), that when implemented, corrects the identified root cause and thereby prevents (or significantly reduces the likelihood of) the incident from recurring.

(Continued)

TABLE 18.1 (*Continued*)
Checklist for Incident Investigations/Reviews

Incident Investigation Elements	Checklist Items	Considerations
	Identify the causal factors related to the incident	• If causal factors cannot be identified, more information may need to be collected or assistance from an Incident Investigator (if your Organization has such a person).
	For each causal factor, determine the root cause(s).	• In consultation with the review team determine root causes • First aid injury investigations are not required to determine root causes.
	Develop corrective actions for the causal factors (for first aids) and root causes (OSHA recordables).	• Review the root causes and consider corrective actions • Corrective Actions should be considered for a lesson learned notice
	Objectives of corrective actions	The corrective actions should have the following properties: • Specific • Measurable • Accountable • Reasonable • Timely • Effective • Reviewed

19 Non-Beam Hazards

Ken Barat

CONTENTS

Non-Beam Hazards are all hazards arising from the presence of a laser system, excluding direct human exposure to direct or scattered laser radiation.

With the very first ANS Z136.1 *Safe Use of Lasers* in 1973 the inclusion of non-beam hazards has been a part of laser safety. Some of this trace back to a high electrical hazard of early power supplies and laser systems. The fact is, as I have said before, laser eyewear does one no good if you are electrocuted.

When one looks at laser incidents, there is a high percentage of non-beam incidents, especially in the industrial setting. This is one of the chief reasons the laser standards insist that non-beam hazards be part of laser standard operating procedures. For that matter, during a risk assessment or safe plan of action review, considerable effort goes to non-beam concerns.

MAGNETIC SAFETY

While one can make up a long list of non-beam concerns, their importance will vary with the work and work environment. An example of this is magnetic safety. First, one may say if I am not working with cryogenic magnetics, or large static magnets then this hazard is of little to no concern. But I have seen individuals injure fingers caught between a screwdriver head and a Faraday Rotator. A Faraday Rotator does present a strong magnetic field at close range. The use of Faraday Rotators might not even make the risk assessment list if one is not familiar with its properties. A Faraday Rotator is a magneto-optic device, where light is transmitted through a transparent medium which is exposed to a magnetic field. The magnetic field lines have approximately the same direction as the beam direction, or the opposite direction. The plane of linearly polarized light is rotated when a magnetic field is applied parallel to the propagation direction (Figure 19.1).

ERGONOMICS

A non-beam hazard that many times goes overlooked is ergonomics. The concern I am talking about is reaching elements on the optical table—un-natural bending of the body and reach in problems. Sometimes this can be overcome by placing platforms around the optical table or where one needs to stand. This, by raising one's height, can usually reach over elements and get to the middle of the optical table.

In the office setting, a great deal of effort and money goes into computer workstations. The opposite can be said of keyboard use and position of monitors in the research laboratory. One excuse for this lack of attention has been the lesser number of hours spent at the keyboard in the lab setting compared to an office setting. While this may be true for many individuals, it is not true for all, and why should we ignore the lab setting (Figure 19.2).

FATIGUE

Fatigue, has two components: excessive workhours and lack of sleep. Many times, excessive work hours are self-imposed. Wise up after about 10 hours on the job as your decision-making process starts to slip. Closely related to that is standing for long periods of time; here, antifatigue mats can be useful. The other component, lack of sleep, is what many researchers suffer from. I cannot understand the boasting remarks of "I can get by with only 4 hours' sleep." Why?

LIMITED WORK SPACE

There is limited work space or area in many laser system installations. Such limited work space can present a problem while working near or around mechanical set-up or high voltage. There should be sufficient room for personnel to turn around and

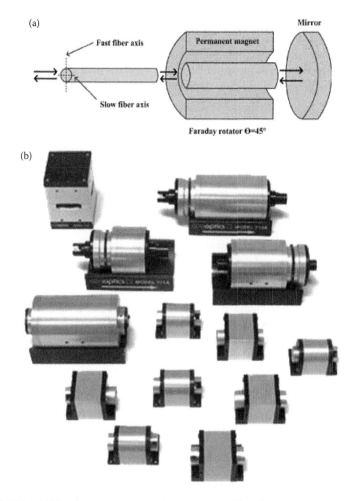

FIGURE 19.1   (a) Faraday rotator set-up, (b) Examples of Faraday rotators. (The photograph was kindly provided by GMP Switzerland.)

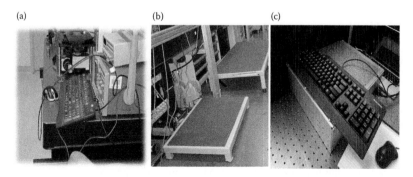

FIGURE 19.2   (a) Keyboard on optical table, (b) Platform of different heights, (c) Keyboard on rack designed unit.

maneuver freely. This issue is further compounded when more than one type of laser is being operated at the same time. The presence of wires and cables on the floor of limited work areas can create trip and slip hazards.

Laser facilities can pose a hazard to laser workers due to obstacles, ambient lighting, confined workplaces, indoor temperature, and humidity.

Preventive Measures

- Provide adequate lighting in the laser-controlled area, or luminescent devices on equipment corners, switch locations, aisles, and so on
- Remove any obstacles inside the nominal hazard zone
- Install the cables, gas tubing, or water hoses in a proper way
- Arrange the laser workplace to have a safe working environment
- Avoid condensation on laser equipment, such as optical components, electrical devices, and so on
- Provide suitable room temperature for the operation of laser equipment

CLASSICAL NON-BEAM CONCERNS

The most classical non-beam hazards are electrical, and laser generated air contaminates. The laser ANSI standards break non-beam hazards into three broad categories: Physical, Chemical, and Biological. The LSO and laser user do not need to be an expert in these areas, but rather be aware that such hazards exist and be alert for them. Once identified, the appropriate safety professionals will need to perform a proper evaluation and advise on control measures mitigating the hazards.

ELECTRICAL ITEMS

Electrical equipment in general presents several potential hazards including: electrocution, resistive heating, ignition of flammable materials, and arc flash. Every institution needs to have an EHS Manual with an Electrical Safety chapter geared to the common electrical hazards their staff might encounter. Some general items to consider:

Electric Shock

The use of lasers or laser systems can present an electric shock hazard. These exposures can occur during laser set-up or installation, maintenance, modification, and service, where equipment protective covers are removed to allow access to active components. A contact with energized electrical conductors contained in device control systems, power supplies, and other components is another way to receive an electric shock. With the use of large power supplies and repetitively pulsed lasers, there is a great potential for electric shock. Shocks usually happen when a person is working on equipment that is not properly grounded or has a large capacitor bank that was not discharged. Most injuries to personnel involving lasers are of this type.

Electric shock is a very serious opportunistic hazard where the occurrence and outcome are difficult to predict, and loss of life has occurred during electrical servicing and testing of laser equipment incorporating high-voltage power supplies.

Protection against accidental contact with energized conductors by means of a barrier system is the primary methodology to prevent electric shock accidents.

The frames, enclosures, and other accessible non-current-carrying metallic parts of laser equipment should be grounded. Grounding should be accomplished by providing a reliable, continuous metallic connection between the part(s) to be grounded. A presence of an "Emergency Power Off" switch will allow the elimination of electrical hazards during emergencies.

Preventive Measures

- Fluids should not be used or placed near the laser system
- The laser system should be labeled with the electrical rating, frequency, and watts
- Proper grounding should be used for metal parts of the laser system
- Assume that all floors are conductive when working with high voltage
- Consider safety devices such as appropriate rubber gloves and insulating mats
- Make sure that the combustible components of the electrical circuit are short circuit tested
- Check that each capacitor is discharged, shorted, and grounded before allowing access to the capacitor area
- Inspect capacitors containers for deformities or leaks
- Avoid wearing rings, metallic watchbands, and other metallic objects when working near high-voltage environment
- Prevent explosions in filament lamps and high-pressure arc lamps
- Inspect regularly the integrity of electrical cords, plugs, and foot pedals
- Only qualified persons authorized to perform service activities should access a laser's internal components
- Do not work alone
- When possible, only use one hand when working on a circuit
- Follow lockout/tag-out procedures when applicable

Resistive Heating

Heating of a conductor due to electric current flow increases with the conductor's resistance. Unchecked and increasing resistive heating can produce excessive heat build-up and potentially damage/corrode system components. Additionally, touching one of these overheated components could result in a thermal burn to the user, maintainer, or servicer. While laser system designers generally provide sufficient cooling for routine operations, equipment should be regularly checked for excessive resistive heating symptoms such as component warping, discoloration, or corrosion, and repaired as needed.

ELECTRIC SPARK IGNITION

Equipment malfunctions can lead to electrical fires. In addition, electrical sparks can serve as an ignition source in the presence of a flammable vapor. Sometimes, when working with devices that are not generally considered to be an ignition source, they may cause smoke or charring without the presence of an actual flame. Under some situations where flammable compounds or substances exist, it may even be possible that smoke could be initiated by Class 3B lasers.

Fire extinguishers designed for electrical fires might be required to be located in close proximity to the lasers. Components in electrical circuits should be evaluated with respect to potential fire hazards. Nonflammable materials must be used for enclosures, barriers, or baffles.

ARC FLASH

An electrical arcing fault can produce an arc flash that includes intense radiant energy, high-temperature air, a high-pressure wave, and high-velocity shrapnel from the electrical apparatus and housing. Causes of arc flash are human error while working on energized electrical equipment, and malfunction due to equipment age, poor maintenance, or poor design. Workers involved in arc flash may incur serious injury or death.

EXAMPLES OF ELECTRICAL INCIDENTS

CASE 1: ELECTRICAL LASER ACCIDENT

A researcher was working on a home-built 5-watt CO_2 laser, replacing and tuning a new tube. The system was obtained from another group. The researcher was unaware that the end caps were energized, and, during the tuning process, he made contact with the palm of his left hand on the end cap and his thumb on the (ground) support bar. He received a painful shock from approximately 15,000 volts at 20 mA DC. The researcher was treated for injury to his hand and wrist, and the pain (severe) lasted several days. Once again, these non-beam hazards can be the deadliest and tend to sneak up on one, for the user is not focusing on them while involved in other activities. This is especially true when working on equipment one is not completely familiar with.

CASE 2: ELECTRICAL SHOCK (REPORTED THROUGH CDRH MAUDE DATABASE)

A service technician was working on a Medlite C6 laser at an office. The laser was reporting error 22, which is "no end of charge" which means the high-voltage (HV) capacitor is not getting fully charged and the flashlamp was not flashing. He evaluated the system and found the simmer supply was working, the lamp was simmering, but the lamps were not flashing. He decided the HV power supply should be replaced. He reports that he turned the system off, unplugged the system, and laid down on the floor to remove the power supply. He disconnected the HV cable from the supply that goes to the HV capacitor, the ac input cable to the supply, and the last thing he remembered was removing the control cable to the supply and then he received the shock. The office staff heard a loud bang from the room, found him bleeding from the ears, nose, and mouth, and they called the building manager who tried to resuscitate him. The emergency response team resuscitated him and took him to the hospital for treatment.

Manufacturers Response

They submitted a report on the site evaluation of the system and interviews with the office personnel and with the service technician. The high-voltage power supply, SCR board and the charge capacitor were returned from the system for further evaluation: the SCR board was intact, the high-voltage discharge relay and discharge resistors were intact and the resistance that was connected to the charge capacitor to discharge the capacitor measured 33 kilo ohms. This was consistent with the design which had 3,100 kilo ohm resistors in parallel. The high-voltage power supply was opened and they found that the bridge rectifier on the input ac was burned and all diodes were shorted, which would not allow the power supply to develop a high-voltage output. This was consistent with report which indicated that when they turned on the system, it reported an error 22 and there was no voltage developed on the charge capacitor. In the report, the service technician indicated when the system would not flash and was reporting an error 22, he also noticed a burning smell. This was consistent with the burned input bridge rectifier in the power supply. The capacitor was received in good condition. There was some evidence that the threads on one of the posts were damaged, not cross threaded, but the post had some rough spots on the threads. This is consistent with the observation in the report that the nut on the wire going from the charge cap to the SRC board was slightly loose. It was observed that you could move the wire on the post, that it was not tight, but it also was not sloppy loose. The diameter of the ring lug on the wire that goes on that post was very close to the diameter of the post so there was little chance that the loose wire could have been really disabled from discharging the capacitor. The backup discharge resistor on the capacitor measured 2.2 mega-ohms which was consistent with the design. None of the evidence identifies defective components, parts, or design that would cause the accident to happen. The only explanation that was possible is the power supply was able to deposit a charge on the capacitor as it was failing. In addition, the loose nut on the capacitor broke the connection to the fast discharge resistors on the SCR board; therefore, the back-up bleed resistor was discharging the partially charged capacitor when the service technician contacted a high-voltage point when he was removing the low-voltage cables from the power supply. The only conclusion is if he had discharged the high-voltage capacitor as he had been recently instructed to do, this accident would not have happened even if one of the system safety discharge circuits was inoperative. Because of the serious nature of the incident, they took additional preventative action steps to impress upon the service personnel the importance of following a specific safety regimen when working in and around high voltage.

Preventative Action

A service bulletin was generated and emailed to all the service engineers and distributors worldwide. This bulletin specifically addressed the safety precautions that must be observed when working in and around high-voltage components. It also stated the minimum wait time that should be observed before attempting to discharge the high-voltage capacitor to allow the back-up bleed resistor to discharge the capacitor to minimize the risk if the capacitor is not fully discharged. The manufacturing and engineering personnel will also be trained in these safety procedures.

LASER GENERATED AIR CONTAMINANTS (LGAC)

Air contaminants may be generated when certain Class 3B and Class 4 laser beams interact with matter. The quantity, composition, and chemical complexity of the LGAC depend greatly upon target material, cover gas, and the beam irradiance. While it is difficult to predict what LGAC may be released in any given interaction situation, it is known that contaminants, including a wide variety of new compounds, can be produced with many types of lasers. When the target irradiance reaches approximately 10^7 W/cm^2, target materials including plastics, composites, metals, and tissues may liberate carcinogenic, toxic, and noxious airborne contaminants. The amount of the LGAC may be greater for lasers that have most of their energy absorbed at the surface of the material.

Some examples include:

- Polycyclic aromatic hydrocarbons from burns on poly (methyl methacrylate) type polymers
- Hydrogen cyanide and benzene from cutting of aromatic polyamide fibers
- Fused silica from cutting quartz
- Heavy metals from etching
- Benzene from cutting polyvinyl chloride
- Cyanide, formaldehyde, and synthetic and natural fibers associated with other processes

Exposure to these contaminants must be controlled to reduce exposure below acceptable OSHA permissible exposure limits. The safety data sheet (SDS) may be consulted to determine exposure information and permissible exposure limits.

In general, there are three preventive measures available: exhaust ventilation, respiratory protection, and isolation of the process. The priority of control requires that engineering controls be used as the primary control measure, with respiratory protection (and other forms of PPE) used as supplementary controls.

- Exhaust Ventilation. Whenever possible, recirculation of plume should be avoided. Exhaust ventilation, including use of fume hoods should be used to control airborne contaminants. Exhaust ventilation systems (including hoods, ducts, air cleaners, and fans) should be designed in accordance with recommendations in the latest revision of ACGIH Industrial Ventilation and ANSI Z9.2: *Fundamentals Governing the Design and Operation of Local Exhaust Systems.*
- Respiratory Protection. Respiratory protection may be used to control brief exposures, or as an interim control measure until other administrative or

engineering controls are implemented. If respiratory protection is utilized, it should comply with the provisions specified in 29 CFR 1910.134.

- Process Isolation. The laser process may be isolated by physical barriers, master-slave manipulators, or remote-control apparatus. This is particularly useful for laser welding or cutting of targets such as plastics, biological material, coated metals, and composite substrates.

The most effective means of reducing the concentration of LGAC is by employing proper ventilation and air filtration systems. Local exhaust ventilation (LEV) systems can effectively capture the air contaminants in close proximity to an emission source. A LEV system is designed to draw contaminated air (LGAC) from a laser process through a partial enclosure or hood at the source of lasing site. The contaminated air is exhausted outside the laser workplace through particulate filtration process. A typical LEV system in an industrial setting employs five major components: a hood or enclosure to capture the LGAC, ducts to carry the contaminated air, suitable fan to provide the airflow, filters to absorb the contaminants in the air, and stack to exhaust the cleaned air.

However, no LEV systems are 100% efficient in capturing all dusts, vapors, and fumes in the air. General ventilation should be provided to reduce the concentration of the air contaminants not removed from the LEV.

Control of LGAC includes, but is not limited to, the following work practices and preventive measures:

- Modify the process or work practice to produce less fume
- Restrict the number of workers present in the hazardous fume zone/laser controlled area
- Limit the duration of exposure
- Perform decontamination prior to taking part in other activities. Washing hands is a good protection of LGAC. A work uniform should be changed daily
- Food, beverage, and cigarettes are not allowed in the NHZ
- Provide respiratory protection training
- If laser generates biological agents, infection control should be part of the training
- Ensure staff uses the PPE properly and effectively
- Implement preventive maintenance schedule for exhaust ventilation systems
- Good communication between management and workers is essential

A professional industrial hygienist should be consulted for the requirements of a ventilation system.

INFECTIOUS MATERIALS

Infectious materials, such as bacterial and viral organisms, may survive beam irradiation and become airborne. To prevent inhaling infectious laser plumes it is recommended the use of high filtration (0.1 micron) masks along with a high-efficiency smoke evacuation system.

Note: Odor, is caused by the toxic gases that are released when tissue pyrolysis and destruction occur as the hot tool impacts the tissue.

The following is a list of chemicals found in the plume from medical laser procedures:

Acrolein	Acetonitrile	Acrylonitrile	Acetylene	Alkyl benzenes
Benzene	Butadiene	Butene	Carbon monoxide	Creosols
Ethane	Ethylene	Formaldehyde	Free radicals	Hydrogen cyanide
Isobutene	Methane	Phenol	Polycyclic aromatic	Hydrocarbons
Propene	Propylene	Pyridine	Pyrrole	Styrene
Toluene	Xylene			

COMPRESSED GASES

Hazardous gases (for example, chlorine, fluorine, hydrogen chloride, and hydrogen fluoride) are used in some laser applications. All compressed gases having a hazardous material information system (HMIS) health, flammability, or reactivity rating of 3 or 4 must be contained in an approved and appropriately exhausted gas cabinet that is alarmed with sensors to indicate potential leakage conditions. Procedures must be developed for safely handling compressed gases (Figure 19.3). Appropriate safety measures should be implemented to avoid the following safety issues associated with compressed gases:

- Working with a freestanding cylinder not isolated from personnel
- Failure to protect open cylinders (regulator disconnected) from atmosphere and contaminants
- No remote shutoff valve or provisions for purging gas before disconnect or reconnect
- Labeled hazardous gas cylinders not maintained in appropriate exhausted enclosures

FIGURE 19.3   Gas cylinder storage.

- Gases of different categories (toxic, corrosive, flammable, oxidizer, inert, high pressure, and cryogenic) not stored separately in accordance with OSHA and Compressed Gas Association requirements

Preventive Measures

- Sensors must be installed in hazardous gas cabinets and other locations as appropriate, including exhaust ventilation ducts.
- Exhaust ductwork should be of rigid construction, especially for hazardous gases.
- Sensors and associated alarm systems should be used for toxic and corrosive chemical agents such as halogen gases. Sensors should always be able to detect the hazardous gas in a mixture of emitted gases (such as, fluorine versus hydrogen fluoride).
- Gas detection systems must be properly shielded to minimize susceptibility to electromagnetic interference (EMI).

CRYOGENIC LIQUIDS

Liquid helium and nitrogen, for example, may be used to cool the laser crystal and associated receiving and transmitting equipment. These liquefied gases are capable of producing skin burns and may replace the oxygen in small unventilated rooms.

Preventive Measures

- The storage and handling of cryogenic liquids must be performed in a safe manner.
- Insulated handling gloves of quick removal type should be worn.
- Clothing should have no pockets or cuffs to catch spilled cryogenics.
- Suitable eye protection must be worn. If a spill occurs on the skin, flood the skin contact area with large quantities of water.
- Adequate ventilation must be present in areas where cryogenic liquids are used.

An overlooked item is proper storage of liquid nitrogen tanks, for example, especially if they are living out in the hallway. They need to be anchored or have the wheels blocked (Figure 19.4).

NANOPARTICLES

The term "nanoparticle" generally refers to particles <100 nm in at least one of its dimensions. The increasingly widespread use of nanomaterials has created concerns about the potential hazards posed by engineered nanoparticles. It appears that small size particles may have higher human risks than larger particles due to their increased reactivity potential. Toxicological studies suggest that some nanomaterials can be transported deep into the lungs and tissues, pass through the blood-brain barrier, or translocate between organs. In addition, the greater surface-to-volume ratio associated with these particles can create greater

(a)   (b)

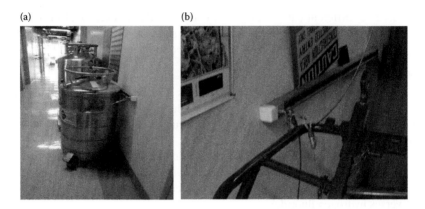

FIGURE 19.4   (a) Cryo tank secured storage, (b) Closer look, earthquake safety.

chemical reactivity than the same material in larger particle sizes, increasing the relative toxicity and fire/explosion hazard presented by a given quantity of material.

Interaction of high energy femtosecond lasers with solid material can cause material blow off (ablation) of fast ions and atoms, as well as clusters and nanoaggregates of target material. The quality and quantity of that energy will determine the amount of ablated material as well as the average particle size. Processes that produce laser-generated nanoparticles must be engineered so as to avoid the entry of such particles into the body via inhalation, ingestion, or absorption processes.

Finally, a potential problem with the production of laser generated nanoparticles is the difficulty of assessing worker exposure and possible subsequent health effects. At the present, there are no occupational standards, appropriate metrics have not been determined, and measurement equipment is not generally available to properly document health effects.

Fiber-Optic Fragment Hazards

Small lengths or particles of optical fiber material can pose a risk of irritation, infection, or injury, particularly when cleaving fibers during splicing operations. All personnel involved with this type of work need to be warned or trained on this hazard. The use of protective finger guards, gloves, or shields should be considered when performing cleaving operations. Adhesive tape can be used to pick up loose particles or splices during work operations. A good work practice is to collect discarded fibers in a suitable container to avoid subsequent embedding in clothing, skin, eyes, or under the fingernails.

Always wear protective eyewear with side shields, even if you normally wear glasses, to prevent any flying shards from getting near your eyes.

Dispose of shards carefully in an appropriately labeled container for sharps. Use a black plastic mat for a work surface. The dark background will make it easier to see

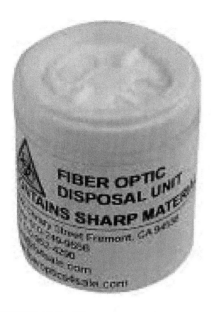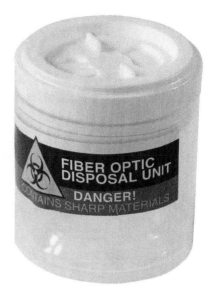

FIGURE 19.5   Fiber scrap containers.

the fibers you are working with and handle them more carefully. Any broken fibers that fall on the mat are easily found for disposal (Figures 19.5 and 19.6).

Preventive Measures

- Always wear safety glasses with side shields and protective gloves. Treat fiber-optic splinters the same as you would glass splinters.
- Keep all food and beverages out of the work area or table where the fibers are cleaved and spliced. If fiber particles are ingested they can cause internal haemorrhaging.
- Wear disposable aprons to minimize fiber particles on your clothing. Fiber particles on your clothing can later get into food, drinks, and/or be ingested by other means.
- Never look directly into the end of fiber cables until you are positive that there is no light source at the other end. Use a fiber-optic power meter to make certain the fiber is dark.
- Do not touch your eyes while working with fiber-optic systems until your hands have been thoroughly washed.
- Put all cut fiber pieces in a properly marked container for disposal.
- Place safety signs up in areas where fiber termination work is being performed.

FIRE HAZARDS

Irradiance levels in excess of 10 W/cm^2 can ignite combustible material. Most Class 4 lasers have irradiance levels exceeding 10 W/cm^2 and are therefore fire hazards.

FIGURE 19.6   Open jar for scrap collection.

Flammable substances can be ignited at even lower irradiance levels making Class 3B lasers possible fire hazards in the presence of flammable substances. Barriers and enclosures around a laser must be capable of withstanding the intensity of the beam for a specific period of time without producing smoke or fire. It is important to obtain information from the manufacturer on the properties of the barrier or enclosure to ensure it will provide adequate protection under worse-case conditions of exposure. Other items such as unprotected wire insulation and plastic tubing can catch on fire if exposed to sufficiently high reflected or scattered beam irradiance. When working with invisible wavelength lasers, this should be kept in mind since it may not be obvious that these surfaces are exposed. The two types of lasers most commonly associated with fires are CO_2 and Nd:YAG lasers. Therefore, provisions must be made to prevent and respond to laser related fires should they occur. The control measures include using noncombustible material in the laser-controlled area especially in the beam path and having adequate fire protection of the facility including sprinkler systems, fire extinguishers, and so on.

Tables 19.1 to 19.4 show a partial list of non-beam hazards and some examples:

TABLE 19.1
Physical Non-Beam Hazards

Hazard	Possible Source
Noise	Constant pinging of pulse laser
Pressure	Vacuum chamber, gas cylinders
Incoherent radiation	Broadband light source
X-rays	Target interaction
High temperature	Ovens in lab
Low temperature	Cryogenic use
Electricity	Power supplies
Trailing cables & pipes	Housekeeping
Sharp edges	Razor blades
Moving parts	Robots
Water- high-pressure	Cooling lines

TABLE 19.2
Chemical Non-Beam Hazards

Hazard	Possible Source
Toxic substances	Laser dyes
Carcinogenic substances	Solvents
Irritant substances	Samples
Dust & particulates	Cracked optics
Fire	From ignition

TABLE 19.3
Biological Non-Beam Hazards

Hazard	Possible Source
Microbiological Organism	From target interaction
Viruses	Released from target interactions

TABLE 19.4
Human Factors

Hazard	Possible Source
Workstation layout	Hitting head on table shelves
Manual handling	Lifting of lasers
Person-machine interface	Robotic work
Shift patterns	Working too many or odd hours

20 Laser Safety Tools

Ken Barat

CONTENTS

When one thinks of laser safety, most often laser protective eyewear comes to mind. But as any LSO knows, there are more laser safety tools than just eyewear. In this chapter, a few items that can be used to block, enclose, or detect beams will be listed for one's possible use in your laser application. When a commercial product is named specifically it will because the product is unique to my knowledge, otherwise generic product terms will be used.

NEWER SOLUTIONS

CARBON RESIGN LIGHTWEIGHT BREAD BOARDS

If weight is a consideration, carbon resin breadboards should be considered. They are more expense that the standard metal breadboard, but weigh 75% less.

BLACK AL FOIL

There are several online sources to obtain these. They were designed for photo studio use, but have found their way to many a laser lab, since they are nonreflective, inexpensive, and can be bent into shape for temporary use.

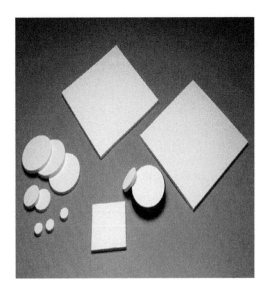

FIGURE 20.1   Spectralon.

DIFFUSE REFLECTION MATERIAL

Spectralon® Diffusion Material gives the highest diffuse reflectance of any known material or coating over the UV-VIS-NIR region of the spectrum. The reflectance is generally >99% over a range from 400 to 1500 nm and >95% from 250 to 2500 nm. The material is also highly Lambertian at wavelengths from 0.257 mm to 10.6 mm, although the material exhibits much lower reflectance at 10.6 mm due to absorbance by the resin. The surface and immediate subsurface of Spectralon exhibits highly Lambertian behavior. The porous network of thermoplastic provides multiple reflections in the first few tenths of a millimeter of Spectralon (Figure 20.1).

INDIRECT LASER BEAM VIEWING TOOLS

LAMINATED IR-VIEWING CARDS

The IR viewing card is designed to allow one to see invisible infrared beams (Figure 20.2). The majority of IR cards found in laser labs to protect the fluorescent material from oxidation are covered with a plastic film. Unfortunately this yields a specular reflector, they are often held by hand (and hence wobbling at all angles) (Figure 20.3). One suggestion is to peel off the coating or use nonlaminated ones. Sensor cards can also be found for ultraviolet wavelengths, but less commonly used. It is important to ALWAYS tilt the IR sensor card DOWN so that any reflections are not directed to yourself nor anyone else standing around.

Sensor cards are not invincible. Know your expected irradiance, as you can burn through these cards and present a fire hazard. As a general rule, NEVER leave an IR sensor card nor any combustible card/plastic/beam block in a beam path unsupervised for an extended period of time.

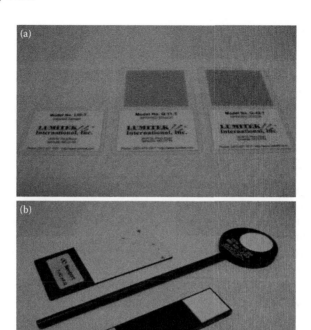

FIGURE 20.2   (a) IR sensor card, example 1, (b) IR sensor card, example 2.

FIGURE 20.3   Hand held IR viewer.

IR Viewers

IR viewers have been a staple in laser labs for decades. The safety concern with these falls into two camps. First, can one use it with their laser protective eyewear on? Depending on your eyewear, the greenish view through a viewer may be difficult viewing, prompting one to take their eyewear off, yielding unprotected eyes.

Second, it's tempting to look at the beam directly, thinking of the viewer as being eyewear. Although a direct beam will not transmit through an IR viewer, a direct beam viewing through an IR viewer will likely have a blinding effect to the eye by overwhelming the sensor, as well as risking damaging the IR viewer.

Neither of these are good or safe practices.

A superior, although more laborious alternative is remote viewing with an IR camera, which removes you from standing in front of the beam or reflection.

HANDS-FREE IR VIEWER

The best of these are really designed as night vision systems and will not claim to be laser detection devices (Figure 20.4). Many will detect out to 1200 nm. Please note, as with any optics, superior optics will cost more. A $2000 night vision goggle will give better resolution than a $400 pair.

A number of commonly used items if not used properly can become a source of hazardous reflections or dangers to the user (Figure 20.5).

CCD/WEBCAM

The common webcam to some iPhone cameras can be used to view visible and NIR beams. The advantage of these devices is that they remove one from the optical table. They come in a number of varieties from commercial to homemade (Figure 20.6).

Combined with the use of motorized mounts, they can make alignment a simple activity.

FIGURE 20.4   Hands-Free IR Viewer set-up.

- REFLECTIVE ALIGNMENT TOOLS: *Beware to ALWAYS consider reflected beam off hemostats and IR card during alignments*

FIGURE 20.5   Reflective objects.

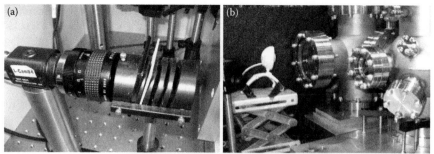

FIGURE 20.6 (a) CCD Camera, (b) Web Cam, (c) Homemade system.

BEAM BLOCKS

Many items fall under the definition of beam block, but not all were designed to be such. Use of a notecard, Post-It Note, or other paper-based items as a beam block is not recommended (Figure 20.7).

While inexpensive, they easily fall over, yielding a suddenly unblocked beam and they sometimes slowly (or not so slowly) burn through if placed at a point where the

FIGURE 20.7 Using paper as beam block.

beam is intense enough—suddenly letting the beam through and maybe extending far down the optical table or off the table. They are commonly found used as blocks for optics transmission, block diffuse reflections, or primary beams.

When using cards or paper as temporary laser shielding, it is important/essential to know which color to choose to avoid laser beam absorption in the card/shield and therefore, risk burning and/or heating. Also note that leaving a card as a block in front of an optic will often out gas and leave residue on the optics which can ultimately damage the optics if not cleaned properly.

Unsecured Beam Blocks

The majority of beam blocks (metal) are designed to be secured to the optical table, either by being screwed down, having a magnetic base, or just using their weight and center of gravity. With beam blocks that cannot be secured, one might say they can be easily moved; they can also be easily moved out of position or knocked down. This type is usually a bent metal sheet or folded cardboard. The range of size and protection of beam blocks vary (Figure 20.8).

Beam Dump

A Beam Dump can be considered a heat sink. It captures a diverted beam. These are either air or water cooled depending on the amount of energy they are indented to deal with (Figures 20.9 and 2.10).

Polycarbonate Sheets

These can be used as beam blocks and perimeter guards for UV and Carbon Dioxide wavelengths. Thus, giving a clear view of the optics on the table.

FIGURE 20.8 Commercial Beam Block.

FIGURE 20.9 Commercial beam dumps.

FIGURE 20.10 Pipe fittings can be used, if no budget exists. Homemade beam dump.

PLASTIC LASER ENCLOSURES

Plastic/acrylic laser enclosures that are rated for certain wavelengths and provide a tested optical density (filtration) can be expensive. Most commonly, people buy plastic or acrylic sheets from a supply catalog (Figure 20.11). Depending on the wavelengths being used, they are effective containment for scatter or direct beams. One of the better designs is ones that have a diffuse film on one side (should be set as the interior not exterior surface of the enclosure). Using a spectrometer and/or power meter, one can self-test the materials. The choice of a proper plastic laser enclosure should never be based just on a visual or "feel good" evaluation. Remember, as one adds new wavelengths to a system, the enclosures containment should be reconfirmed.

METAL LASER ENCLOSURES, TABLE PERIMETER GUARDS

These types would seem to answer the uncertainties listed above for plastic enclosures. The cautions with metal enclosures is burning off the coatings, and making sure it does not present a specular reflection source (Figure 20.12).

(a)

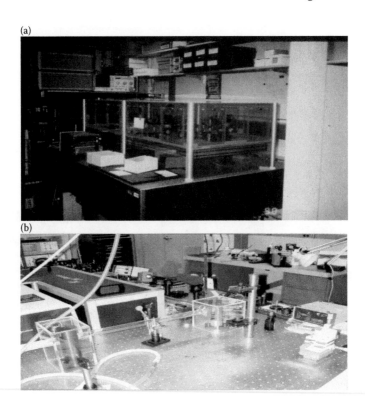

(b)

FIGURE 20.11   (a) Wavelength specific acrylic enclosure, (b) Polycarbonate enclosures, UV wavelengths.

LASER CURTAINS

Laser curtain are most commonly used to segregate areas of a laser lab. In general, eyewear required zones from not required zones. Unless for required lighting conditions, laser curtains should not be floor to ceiling. At ceiling height, they interfere with the water distribution pattern of fire suppression sprinklers. Meaning the sprinkler heads need to be lowered to be effective. Laser curtains can be certified laser curtains or in some cases opaque welding curtains, contact the LSO for options. There is a considerable price and performance difference between the two. Laser curtains can also be made of metal (Figure 20.13).

LASER PROTECTIVE EYEWEAR

Laser protective eyewear is one's last line of defense against laser beam exposure. This is discussed in detail in Chapter 10.

There are more things out there that can have laser safety applications, however, the ones mentioned in this chapter are just to get you thinking. Kentek, Thor Labs, Lasermet, Newport, and Laservision are all good sources for items that can help make the laser user safer.

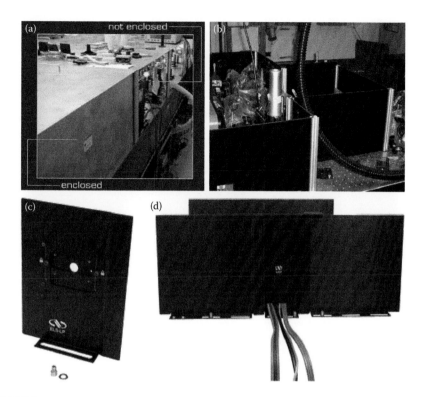

FIGURE 20.12 Newer style that allows cables and tubes to leave the table while still providing a high level of protection. (a) Table enclosure, practical, (b) Perimeter guards, (c) Perimeter guard, cable opening, (d) Perimeter guard, cable exit layout.

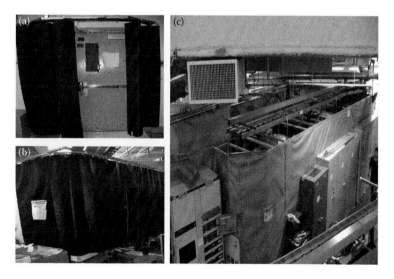

FIGURE 20.13   (a) Interior curtain, laser free zone, (b) Laser curtain, example 1, (c) Laser curtain, example 2.

21   How Many Wrong Decisions Can One Rationalize Away?

Ken Barat

CONTENTS

The follow is a laser incident that highlights a long list of actions that were all wrong.

SUMMARY OF INCIDENT

A graduate student was working with a researcher testing a new linear antenna design to determine its sensitivity. The task involved testing silicon samples to

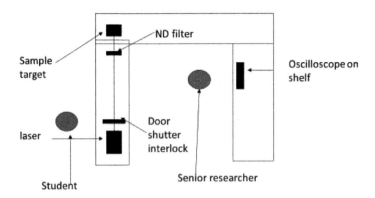

FIGURE 21.1 Lab layout.

compare the amplitude of signals produced by the new design antenna versus the existing antenna. This experiment had been going on for 1 year, without any incidents.

The experimental set-up was designed for two people, one to operate the laser and handle the sample, the other to be positioned near an oscilloscope to observe the effects of the laser on the sample. Figure 21.1 shows the lab layout.

The researcher left the room to obtain a new sample. The experimental layout contained a neutral density filter that attenuated the output of the laser onto the sample. Once alone, the student decided to remove the neutral density (ND) filter to increase the output striking sample and to see the outcome. In order to see the oscilloscope from his side of the lab, the student needed to remove his eyewear or walk to the other side of the lab (do I need to say which choice he made).

Removal of the ND filter had no effect on readings, so using tweezers (stainless steel), the student removed the sample. At that time, he observed a multicolored flash in his right eye. At the time, he thought the flash was from the room light. Laser in use was an Nd:YAG laser pumping an OPO, output wavelength 750 nm at 22 mJ/pulse at 10 Hz, with ND filter in use, output was reduced to 22 uJ/pulse. A barely visible beam would be present when the laser was on.

When the researcher reentered the room, he noticed the student not wearing any eyewear and instructed him to put it back on. A few minutes later, the student noticed fuzzy vision in his right eye, then a dark spot that obscured the vision in his right eye. He told the researcher, who shut down the laser and took him to medical. A medical exam confirmed the student had received a retinal exposure.

FACTORS TO CONSIDER

Here is more background information on this lab incident, and each action deserves a comment (Figures 21.2–21.4).

FIGURE 21.2   Lab set-up.

FIGURE 21.3   Closer look at the scope.

Researcher Left the Room to Obtain a New Sample, Once Alone, Student Removed ND Filter to Increase Output Striking Sample

While the student's decision may have been based on scientific curiosity, the beam output went from 22 microjoules to 22 millijoules. Would the student's eyewear provide protection at this level?

Some have suggested the filter should have been locked in place to stop removal.

To See the Oscilloscope, the Student Removed His Eyewear. Oscilloscope on Opposite Optical Bench Displays Experimental Results

The question to ask is why was the scope so far from the student, could have the scope been moved to a better location? Maybe a camera could have been aimed at the scope and a monitor positioned near the student's work area. Experimental layout is questionable.

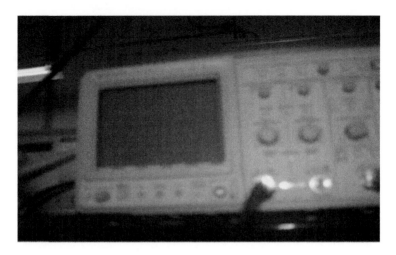

FIGURE 21.4 View of scope with eyewear on.

Lab Had Not Been Visited by LSO for Over 2 Years

The LSO position was only a small percentage of the LSO's responsibility. Not being familiar with laser work, the LSO left laser safety primarily to users.

No Laser Warning Signs Used on Doors

While the absence of a warning sign has no effect on how the student was injured, it does demonstrate the lack of attention laser safety received.

Standard Operating Procedures (SOPs) for This Lab
Were Completely Generic in Content

Once again, this demonstrates the lack of attention laser safety was receiving. While it is easy to write generic SOPs, the need to have site-specific items cannot be overstated. A procedure on how to change samples might have highlighted blocking the beam. Remember when the LSO signs the document, it is an indication that laser safety is adequately being addressed.

Removal of the ND Filter Had No Effect on Readings, So Using
Tweezers (Stainless Steel), the Student Removed the Sample

If the laser could not be shut off for thermal stability reasons, then why was the beam entering the sample box not blocked? A beam dump or power meter or diverting the beam would have been an easy approach to this concern. Giving a laser-free area and keeping the laser on. If we turn to the tweezers, why stainless steel, why not plastic or anodized tweezers to reduce the possibility of specular reflections.

At That Time, He Observed a Multicolored Flash in His Right Eye.
At the Time, He Thought the Flash was from the Room Light

At no time during the investigation was the question asked of why the student thought the flash was a flicker of the room lights. Of course, the flash may have been a routine item, the source of the student's eyewear being hit from a reflection off the tweezer.

Emission Indicator on Laser Was Broken and Did
Not Indicate If the Laser Was On or Off

There is no question that the student knew the laser was on. It was his responsibility to operate the laser. Therefore, the broken emission indicator (usually just an LED) is an example of poor care and disregard for all but the collection of data that hung over the lab.

QUESTIONS YOU SHOULD BE ASKING

Why Was the Beam Path So Long?

This was not a pulse probe set-up, the entire beam path could have been shortened.

Why Was the Beam Path Open?

There was nothing that would have negativity affected the set-up if the beam was in a beam tube or was in an enclosure.

Did the Student Receive Any Laser Safety Training?

During the investigation that followed the incident, several individuals claimed that the student had received basic/institutional laser safety training, but no documented record could be found to support this.

Did They Have the Correct Eyewear?

Two pairs of laser protective eyewear were found in the lab (Figure 21.5). One was labeled for the NIR wavelength in use. Unfortunately, the plastic filter dye made

FIGURE 21.5   Eyewear used in lab.

reading the oscilloscope from any distance impossible. This is the pair given to the student to use; it was properly labeled. The other pair was clear and made of glass, but not labeled for the wavelength in use (this pair was used by the researcher). When challenged on this, the researcher indicated contact with the eyewear manufacturer indicated this pair would provide adequate protection from the attenuated beam, 22 uJ. Of course, when looking at the lab set-up, about 75% of the beam path was at the higher output.

Housekeeping

From images of the lab, it appears that housekeeping, such as the warning sign, seemed to be missing or at least lacking, from the optical tables to cable management.

THE MOST SERIOUS PROBLEM OR CONCERN

THE EXPERIMENT HAD BEEN OPERATING FOR OVER 1 YEAR

What does this mean? My interpretation is that the experiment was set-up, got to work, and then they were off to the races. At no time did anyone ask: Could we make this set-up better let alone safer? In less than 2 days, this set-up could have been easily improved. Going from a danger zone to a safe and efficient experiment.

SUMMATION

This incident contains a large collection of laser safety errors. So many, one might think this case is not real, but sad to say, it really happened. As stated, if they had stepped back and asked, "What can we do to make things safer and easier on ourselves," this incident most likely would never had happened.

Index

Note: Page numbers followed by "*fn*" indicate footnotes.

Printed and bound by CPI Group (UK) Ltd, Croydon, CR0 4YY

23/10/2024

01777951-0001